Easy Things to See With a Small Telescope

A Beginner's Guide to Over 60 Easy-to-Find Night Sky Sights

Cover Image: Gulliame Blanchard, French optical engineer is photographing (from left to right): Jupiter, Venus, Moon conjunction from the VISTA (the Visible and Infrared Survey Telescope for Astronomy) site; on his left is Laura Ventura.

Credit: ESO/Max Alexander. Used under the Creative Commons Attribution 4.0 International License.

Original Image: http://www.eso.org/public/images/max_v558442s/

Copyright © 2015 by Richard J. Bartlett

All rights reserved. Except for brief excerpts used in a review, this book or any portion thereof may not be reproduced or used in any manner whatsoever without the express written permission of the publisher.

First Edition, December 2015

For Joy

>*The Luna in my night. Thank you for illuminating my life and for staring at the stars with me*

For James

>*Your star rose in my sky nine years ago and has continued to shine brightly ever since. I still steer my ship by its light.*

For Maya

>*Your star is rising and glitters and glows like gold. I can't wait to see how brightly you'll shine in the years to come.*

For Charlie

>*Your star has yet to rise but will be visible in just a few short weeks. I can't wait to wonder at its light.*

Contents

Introduction 7
Telescope Basics 13
Tips 'n Tricks 19
Signposts to the Stars 29
Star Charts & Observing Lists 33
The Things You Can See 59
The Deep Sky Objects 69
Observation Logs 135
Appendix 159

Introduction

Why I Wrote This Book

I've had an interest in astronomy since I was six years old, but I didn't start to use a telescope until I was about nine or ten. My brother had a small, department store telescope that I'd "borrow" on clear nights. I distinctly remember having my bedroom window open, with the little telescope perched upon the sill, staring at the Moon, planets and bright stars as they rose over the trees at the bottom of our garden.

I'd learnt the stars from reading books, like this one. I had an idea of what to look at and although the little telescope was really not much better than a toy, it still provided me with enough inspiration to continue the hobby. A few years later, my parents bought me my first "real" telescope – a small refractor – for my eleven birthday. Suddenly, it was as though the universe had opened up.

Now, besides the Moon, planets and bright stars, I could look much further afield. The Andromeda Galaxy, the Orion Nebula, the Pleiades… even double stars like the wonderful Albireo… they all quickly became favorite targets for my telescope.

But then something happened. The books I owned had wonderful, beautiful images of far-flung night sky sights but either I couldn't find them or they were beyond the reach of my small telescope. Still others remained invisible to me because I simply didn't know they existed. The books I owned – mostly written for young children – only highlighted a handful of very easy objects and I'd already found them.

I still went outside when the skies were clear. I still looked at the Moon, the planets and the night sky sights I knew about. But pretty soon I found myself wishing for more – but without knowing what to look for, I'd simply look at the old, familiar sights and then go back inside.

This book is really for my twelve year old self. The one who got excited about exploring the universe and discovering the hidden treasures of the night sky. The one who wished he knew what else to look for and where to look for it.

If I could go back in time and give him an astronomy book to read, it would be this one.

About this Book

Fortunately, you don't have to be twelve years old to benefit from this book. I've written it for anyone who has a small telescope and who wants to explore the night sky, regardless of their age and experience. My hope is that you'll be inspired to go deeper into the night and discover more of what's "out there."

As you may have already noticed, I'm not really one to use a formal writing style. I'd rather write as though I'm standing beside you and talking to you instead. Likewise, I try to avoid jargon and long pages of explanatory text but, to some extent, that's a little unavoidable. However, if you already own a telescope, know your way around the sky and want to jump right into it, feel free to skip ahead to the star charts on page 33. You don't have to read every word I've written!

If, on the other hand, you're looking to learn a little more (or it's cloudy outside) then feel free to read the other sections too.

The book is divided up as follows:

- Telescope Basics
- Tips 'n Tricks
- Signposts to the Stars
- The Star Charts & Observing Lists
- The Things You Can See
- The Deep Sky Objects
- Observation Logs
- Appendix

In the *Telescope Basics* section, I'll talk a little about the telescopes on the market and which ones might be best suited for you. The *Tips 'n Tricks* section will give you exactly that – a few pointers and handy hints I've picked up over the years. You'll find that many of these are already in practice by amateur astronomers (like yourself!) all over the world.

Signposts to the Stars is a short section about the two most important constellations for beginners: Ursa Major and Orion. More specifically, it'll show you how you can use these constellations to find others in the sky. If you're very new to astronomy, you might find this useful.

The *Star Charts & Observing Lists* section is where the fun begins. The section begins with a table; simply look for the current time of year (for example, early November) and then look for the current time of night (for example, 10 p.m.) The table will then tell you which star chart to use. (In this example, that would be Chart 7.)

The chart will show you the night sky as it appears at that time and also includes a list of objects to observe. The list runs from west (setting) to east (rising) but it's not an all-inclusive list as I've focused on the objects that should be highest over the horizon at that time. Some may be too low to make any observation worthwhile.

In *The Things You Can See* I'll discuss the different objects you'll be able to observe (including the planets) and review the information to be found on the individual pages for the objects themselves.

Next you'll find the biggest section – *The Deep Sky Objects* themselves. Here you'll find information on over sixty "deep sky" objects. For each object you'll find a general star map, a simulated view through a finderscope and a simulation of what you might see through your eyepiece. (See *The Things You Can See* section for more information.)

I always recommend that folks keep a record of their observations (especially if you're new to the hobby) and have included some pages in the *Observation Logs* for you to make some basic notes. There are enough pages for you to make an observation of each of the deep sky objects contained in this book.

Lastly, there's an *Appendix* that contains some useful supplemental information: for example, a basic glossary and recommended resources such as books, software and Facebook groups.

About the Author

Photo by James Bartlett

I've had an interest in astronomy since I was six and although my interest has waxed and waned like the Moon, I've always felt compelled to stop and stare at the stars.

In the late 90's, I discovered the booming frontier of the internet, and like a settler in the Midwest, I quickly staked my claim on it. I started to build a (now-defunct) website called *StarLore*. It was designed to be an online resource for amateur astronomers who wanted to know more about the constellations - and all the stars and deep sky objects to be found within them. It was quite an undertaking.

After the website was featured in the February 2001 edition of *Sky & Telescope* magazine, I began reviewing astronomical websites and software for their rival, *Astronomy*. This was something of a dream come true; I'd been reading the magazine since I was a kid and now my name was regularly appearing in it.

Unfortunately, a financial downturn forced my monthly column to be cut after a few years but I'll always be grateful for the chance to write for the world's best-selling astronomy magazine.

I emigrated from England to the United States in 2004 and spent three years under relatively clear, dark skies in Oklahoma. I then relocated to Kentucky in 2008 and then California in 2013. I now live in the suburbs of Los Angeles; not the most ideal location for astronomy, but there are still a number of naked eye events that are easily visible on any given night.

Also by the Author...

2016 An Astronomical Year is written for everyone with an interest in astronomy and contains information on hundreds of night sky events throughout the year. It was designed for astronomers of all levels and includes details of the lunar phases and eclipses, as well as conjunctions, oppositions, magnitude and apparent diameter changes for the planets and major asteroids.

To date, the 2015 edition has been downloaded nearly 3,000 times, was ranked #1 in Free Kindle Astronomy books, within the Top 10 Paid Kindle Astronomy books and within the Top 50 Free Kindle Non-Fiction books.

It is available in paperback and Kindle editions in the United States, Canada and the United Kingdom. (Please be aware that due to the cost of printing in color, the paperback does not contain images and is purely text only.)

2016 The Night Sky Sights is specifically designed for absolute beginners and casual stargazers without a telescope. The guide highlights over 125 astronomical events in 2016 - all of them visible with just your eyes - and showcases events visible in both the evening and pre-dawn sky as well as those you can see throughout the night.

It is currently available in paperback and Kindle editions in the United States, Canada and the United Kingdom.

The Astronomical Almanac (2016-2020): A Comprehensive Guide to Night Sky Events provides details of thousands of astronomical events from 2016 to the end of 2020. Designed for more experience astronomers, this the guide includes almost daily data and information on the Moon and planets, as well as Pluto, Ceres, Pallas, Juno and Vesta.

To date, the 2015-2019 edition has been downloaded nearly 6,000 times, was ranked #1 in the Free Kindle Astronomy book category, #3 in the Paid Kindle Astronomy book category and within the Top 50 of *all* Free Kindle books in October 2014.

It is available in paperback and Kindle editions worldwide, including the United States, Canada, the United Kingdom and Australia.

The Amateur Astronomer's Notebook: A Journal for Recording and Sketching Astronomical Observations is the perfect way to log your observations of the Moon, stars, planets and deep sky objects. It is available as both a full-size 8.5" by 11" journal and also as a 5" by 8" pocket notebook. The larger edition has room for 150 observing sessions while the pocket edition allows you to record 100 observations.

It is available as a paperback in selected areas. (Full Size Edition: United States, Canada and the United Kingdom. Pocket Edition: United States, Canada and the United Kingdom.)

The Deep Sky Observer's Guide offers you the night sky at your fingertips. As an amateur astronomer, you want to know what's up tonight and you don't always have the time to plan ahead. Maybe the clouds have suddenly parted. Maybe you're at a star party. Maybe you want to challenge yourself with something new but don't know where to start.

The guide can solve these problems in a conveniently sized paperback that easily fits in your back pocket. Take it outside and let the guide suggest any one of over 1,300 deep sky objects, all visible with a small telescope and many accessible via binoculars.

Echoes of Earth – a collection of science fiction, mythological and philosophical short stories that I wrote many, many moons ago. (i.e., in the mid 1990's.)

It is available as a Kindle edition in selected areas. (United States, Canada, the United Kingdom and Australia.)

The Author Online

Email: astronomywriter@gmail.com

Amazon US: http://tinyurl.com/rjbamazon-us

Amazon UK: http://tinyurl.com/rjbamazon-uk

The Astronomical Year: http://tinyurl.com/theastroyear

Facebook: http://tinyurl.com/rjbfacebook

Twitter (@astronomywriter): http://tinyurl.com/rjbtwitter

Clear skies,

Richard J. Bartlett

November 22nd, 2015

Telescope Basics

Refractors vs Reflectors

There are many different types of telescope but they broadly fall into one of two categories: refractors or reflectors.

A refractor is what many people consider to be the classic concept of a telescope. It's basically a tube with a lens at one end and an eyepiece at the other. It sits atop a tripod and is usually the kind found in department and toy stores as a "first telescope" for children.

By the way, there are only a small number of telescope manufacturers I'd consider and, out of those, I've only ever bought from Orion and Celestron. There are others, but both these manufacturers are very well respected in the industry and both produce fine equipment at a reasonable price. If you're looking at a refractor telescope in a department store and it's *not* made by Orion or Celestron, I'd walk away. You might think it's a good buy for the price, but it's probably cheap for a reason.

As you're shopping around, you might come across a telescope that's listed as, for example, a 70mm refractor. This means the lens of the telescope is 70mm in diameter – or just under three inches wide. The size of the lens will dictate how much light is gathered by the telescope and, consequently, how much you'll be able to see as a result.

The 70mm Celestron TravelScope. Image by the author.

Again, stay away from department store telescopes that make ridiculous claims like "magnifies up to 1,000x!" In theory, any telescope can magnify by any amount but the view is worthless because the optics simply won't produce a good quality view. If your optics are imperfect then any imperfections are also magnified by 1,000x.

Also, it's almost a waste of time to try and magnify anything that much. I suspect a lot of inexperienced folk believe that you need a high magnification to see anything – because, after all, the stars and planets are only tiny points of light and those same folks are unaware of everything else you can see at low power.

In reality, only planets need a high magnification and you can still see a lot at 100x, which is well within the range of a decent small 'scope made by a reputable manufacturer.

A good refractor is able to produce a relatively high magnification, which makes it an excellent choice for observing the planets. A refractor also has the benefit of being quite portable as, for technical reasons I won't go into here (I promise to restrict the jargon) the size of the lenses typically range from about 70mm to 100mm in diameter.

Both Orion and Celestron produce some excellent "first 'scope" refractors – especially if you're looking for something portable or you're on a budget. As of this date (November 2015) there are a number of options under $100, such as the Orion 10034 GoScope II or the Celestron 21035 70mm Travel Scope. Both of these come with a backpack that allow you to easily pack up your 'scope and take it wherever you want to go – great for family vacations!

(I personally own the Celestron and have used it under the dark skies of Ojai, California. Talk about reliving childhood memories...)

The other kind of telescope is a reflector. There are, in fact, many different types of reflectors but I'll try to keep this simple and straightforward. Whereas a refractor works by having the light pass through a lens, a reflector works by literally reflecting the light with mirrors.

The light enters through the open end of the telescope tube, called the aperture. It then strikes the large mirror at the bottom of the tube, called the primary mirror, and is reflected back up toward the aperture. Before it can escape again, it hits a much smaller mirror (called the secondary) and is reflected out the side of the tube via the eyepiece.

There's one big advantage to this: a refractor typically requires a long tube because the light isn't reflected and must travel in a straight line. With a reflector, the light is bounced twice and so the required tube length is halved. This has the added benefit of allowing for much larger mirrors and, hence, more light gathering power.

There's one more big difference between a refractor and a reflector. A refractor telescope is almost always found on top of a tripod and there are many reflectors that are similarly mounted. However, there's one very popular mount that's in very common usage today: the Dobsonian.

A Dobsonian mount is relatively simple. The tube sits upon the mount, which, in turn, sits upon a base on the ground. The mount then turns from side to side and allows the telescope to move up and down.

Dobsonian mounts are designed to carry some pretty large beasts and can be permanently grounded. This makes them a favorite with amateur telescope makers who aren't so concerned with portability but would rather have a powerful telescope instead.

(Having said that, you can also have Dobsonians that are relatively small but are still portable, such as Orion's range of SkyQuest XT 'scopes. I owned an XT 4.5 – as seen in this image - and loved it.)

Dobsonians (aka "Dobs") are "light buckets" and are great for deep sky observing but they have one big potential downside – depending on your point of view. Dobs typically aren't motorized.

What does this mean?

Well, when you're looking at an object in the night sky through a telescope, you'll notice that it appears to drift across the field of view as the sky turns above you. Yes, that's right, you will literally see the stars move from east to west.

Having a motorized mount allows your telescope to compensate for this and, consequently, the object you're observing will stay locked within your field of view. This can be critical if you want to try your hand at astrophotography or want to make a detailed sketch.

Reprinted with permission from Orion Telescopes & Binoculars, www.telescope.com

(There are some exceptions to this. For example, Orion produces a range of motorized Dobsonian 'scopes that will guide you to your object.)

GoTo 'Scopes

The Celestron 130 SLT GoTo. Image by the author.

One more thing... technology has evolved a lot over the past few decades and amateur astronomers have certainly benefitted as a result. One of the advancements has been the development of the "GoTo" telescope.

A GoTo is typically a reflector on a motorized mount that can automatically find objects for you through the use of a computerized hand controller. I think I see a see a light bulb appearing over your head, but before you get too excited, I feel I must tell you this:

Although GoTo's are great - I own one – there's *nothing* that can beat tracking down and "discovering" an object for yourself. A GoTo may be able to place an object in your field of view within seconds, but it's nothing compared to the thrill of using your observational skill (and a decent star chart) to track it down yourself.

So, while all of the objects in this book can easily be found with a GoTo, I *strongly* encourage you to go old school and leave the GoTo behind.

I know what you're thinking though... if I'm not recommending a GoTo, why do I own one myself? Well, there's two main reasons. Firstly, and primarily, it's because of light pollution. I live in the suburbs of Los Angeles, where only the brightest of the bright stars can be seen on any clear night. It's extremely difficult to find any deep sky objects under these conditions but the GoTo can find them for me.

Secondly, when I bought my 'scope I wanted something with a motorized mount and they're pretty much all GoTo's now. True, you don't have to use that functionality, but when you have a database of thousands of objects at your fingertips, it's kinda hard to resist it.

(Incidentally, I'd argue that whether you're using a GoTo or not, you still need to know what you're looking for. A GoTo mount will often point the telescope very close to the object but it will rarely place it smack in the middle of your field of view. Sometimes, if you're unlucky, you have to slew the 'scope a little to see your object and if you're not sure what you're looking for, it can be easily missed.)

Which Telescope Should You Buy?

I know I've thrown a lot of information at you and you're probably scratching your head right about now. Don't worry, it gets easier. If you've already bought your telescope, you might even be wondering if you've bought the right one.

In fact, let me address that issue right now. The answer to the question "did I buy the right telescope?" is always "yes. Yes you did." Why? Because anything that allows you to turn your eyes to the stars and to explore the universe is a gift – even if you bought it yourself. When Galileo discovered the four largest moons of Jupiter in 1610 he did it with a simple telescope that makes those department store 'scopes I warned you about look like the Hubble Space Telescope.

So if you bought your own telescope (even from a department store) and you're wondering if you made the right choice, then don't worry about it. If, on the other hand, you haven't bought one yet (or are looking to buy another – congratulations, you've caught the bug) then here are some recommendations.

Under $100 and/or Portable

The Celestron 76mm FirstScope. Image by the author.

You've got some excellent choices for small 'scopes. If you're looking for something as a beginner 'scope, maybe for a child or a young adult, I'd go with either the Orion 10033 FunScope 76mm TableTop Reflector or the Celestron 76mm FirstScope. Both are available for about $70 and can provide great views of the Moon, planets and deep sky objects to get you started. (Almost all of the objects in this book should be within reach of these 'scopes, although that will also depend upon your viewing location and your own eyesight.)

From a parent's point of view, if you buy one of these for little Jimmy and he loses interest after a while (I can't imagine why, but who knows) then you haven't lost much. And as an added bonus, these 'scopes are small and very portable. The only downside is that you'll need something like a table to put them on.

I've personally bought the Celestron FirstScope for my nine year old son and for my girlfriend's five year old daughter; some of the notes in this book are based upon observations made with this 'scope.

Alternatively, you can go with either the Orion 10034 GoScope II or the Celestron 21035 Travel Scope I mentioned earlier. Although they both have a slightly smaller lens (70mm) they're attached to a tripod that can easily be adjusted to suit the observer's height – so no need for a table then.

Under $300

If you're more serious, I highly recommend the Orion SkyQuest XT 4.5 Dobsonian. This is the telescope I owned when I lived in Oklahoma and I loved it. I spent almost every moonless weekend night outside and was able to "discover" something new every night. I've observed almost every object in this book with that 'scope and the only reason why I didn't observe the rest was because I never got around to it. Many of the notes are based upon my observations during that time.

After Oklahoma, I moved to Kentucky where the light pollution from Louisville dampened my spirits. Then nearly six years later, I moved to Los Angeles. I drove from Kentucky to California with all my worldly possessions in my car and, unfortunately, I didn't have room for the 'scope. Consequently, I very regrettably gave it away. In fact, Dawn, if you still have it, you've no excuse for not using it now!

As the name implies, it's a reflector with a mirror that's four and a half inches wide, or about 112mm. It has a handle and weighs about sixteen pounds, so it's portable, but you'll need a small table to put it on. (I used a folding table I bought from a hardware store for about $20 and that worked just fine.)

Under $400

If you're looking for a GoTo – I won't judge – then I recommend the Celestron 130 SLT. This is the 'scope I bought (and still own) after I moved to Los Angeles and, as I said earlier, it's proven to be invaluable under these light polluted skies. The only drawback that is that it's a little large to be taken on a trip. I'd certainly move it for an astronomy weekend, but for a regular family trip I'd take the Travel Scope or the FirstScope instead.

The mirror is 130mm wide, which is just slightly more than five inches. Every one of the objects in this book should be easily visible with this 'scope.

Tips 'n Tricks

Get to Know the Stars

I can't emphasize this enough. Maybe you're already familiar with the night sky and all the constellations, but if you're not, I'm sorry, but there's no way around this. If you don't know your way around the stars then trying to find some of the hidden treasures will be like starting from a random point in the U.S. and trying to get to Disneyland. One easy way to learn the stars is to use a planisphere, like the one below.

Get a Planisphere

I've provided some star charts to help you find your way around the sky but there's something else you can buy which, I believe, is a huge asset to any astronomer. A planisphere. It's basically a disc with the stars plotted upon it that allows you to "dial up" a view of the sky for any night at any time.

It works like this. There are two discs, one attached to the other. On the lower disc is a map of the entire night sky with all the stars that can be seen from your location. Around the edges are the months with marks to indicate the individual dates.

This planisphere is actually small enough to fit inside the toolbox I use to store my eyepieces. Image by the author.

The upper disc is basically a circular mask that covers the sky chart and only allows for a segment of the sky to be visible. The edges of the disc are marked with the twenty-four hours of the day, from midnight to midday and through to midnight again. In order to see a depiction of the night sky on a particular date and at a specific time, you simply move the upper disc so that the desired date and time are aligned. The upper disc will then allow you to see the visible stars and will mask the rest.

I know what you're thinking (and probably saying) – "oh I have an app for that." Yes, you probably do (I have several) – but here's why I don't use the app when I'm observing: firstly, you're relying on your cell phone or tablet battery to keep you going. And if that battery dies and you need your cell phone for the drive home from a dark sky site, you have a potential problem.

Secondly, your eyes need to adjust to the dark (see below) and although many apps will dim the screen, if you hit the wrong button on your device you might find yourself staring at a bright screen instead. And now you've just lost your night vision.

You're probably also saying "how am I supposed to see the planisphere in the dark?" Well, that's what a red flashlight is for (see below.) Ideally, get a planisphere with a white background and black stars (like the star charts I've printed in this book.) This will make it easier for you to see it in the dark.

You can buy a planisphere online from almost any telescope manufacturer, including Orion. Be sure to get the correct one for your latitude otherwise you might still be lost!

Get a Red Flashlight

Image by the author.

A red flashlight is an essential tool for any astronomer. Your eyes need time to adjust to the dark (see below) and if you use a regular flashlight, your night vision will be ruined. On top of that, if you're out with a group and you use a regular flashlight, you run the risk of ruining your fellow astronomer's night vision.

That will make them angry. And you don't want to make an amateur astronomer angry.

All that being said, your eyes aren't adversely affected by red light and, many moons ago, astronomers would buy some clear plastic red tape to cover their flashlights. Nowadays you don't have to go to all that bother as a red flashlight is pretty inexpensive.

You can easily buy them online and may even be able to buy them from stores selling camping and hunting equipment. If you can, get one that allows you to adjust the brightness of the beam. This will make it easier for you to view your planisphere, star charts or even (hopefully!) read this book!

Let Your Eyes Adjust to the Dark

Many people are aware that their eyes grow accustomed to the dark. Stand in a lit bedroom at night, with the curtains drawn and then turn out the lights. What do you see? Probably not much if the room is properly dark.

Likewise, if you need to get up in the middle of the night and you turn the light on, what happens? How do your eyes feel? More than likely you'll be dazzled by the sudden burst of light.

It takes your eyes about an hour to fully adjust to the dark and you don't want to lose that sensitivity (hence the red flashlight.) Without that sensitivity you may not be able to see the finer details in some of the deep sky objects (such as nebulae and galaxies) and some of those objects may not be seen at all.

So what do you do while you're waiting? Well, I for one don't wait around. Take some time to enjoy the night sky. If there's a crescent Moon in the sky, I might take a look at that (but with the use of a filter – see below.) *Don't* observe the Moon after your eyes have adapted to the dark as its brightness will probably ruin your night vision again.

Look at the planets, if any are visible. Multiple stars are also a good choice as their light is concentrated in a single, bright point and most are easily seen. (There are a few examples of faint multiple stars but usually you're good to go.)

As my eyes adjust, I also take some time to look at the brighter star clusters, such as the Pleiades or the Praesepe. These are often comprised of many bright stars and can be a very welcome distraction as you wait for your night vision to kick in. I *do* also recommend you revisit them later in the session as you'll frequently see many more of the fainter stars once your eyes have adjusted to the dark.

Beware the Moon

You may notice that I don't talk about the Moon in this book. There's a reason for this. As pretty as it is, the Moon (like an overcast sky) is not my friend. The Moon brightens the sky, just as the Sun does, so instead of the nice, dark sky you'll get on a moonless night, you'll be looking at a deep blue sky instead.

How does this affect your observing? Well, some of the fainter fuzzies (galaxies and nebulae, specifically) will be lost against the glow of the sky. On the plus side, multiple stars are often unaffected as, again, their light is concentrated into a single bright point that can easily shine through.

I don't typically observe when the Moon is out, or, if it's a crescent Moon I'll observe if it's relatively close to the horizon (as a crescent Moon often is.) Consequently, I'll

The waxing crescent Moon, May 30th 2006. Taken with an Orion SkyQuest XT 4.5. Image by the author.

often wait until about three or four days after the Moon has turned full. By that time, the Moon won't be rising until some time after sunset, giving me the evening hours to conduct my observing. It stays that way until about two or three days after new Moon, when the Moon sets a few hours after the Sun and I can use the rest of the evening for my observing.

Anything between an evening crescent and just after full Moon is a problem because the Moon is visible for much of the evening.

That's not to say the Moon isn't worth observing, because it is – honestly, it can be stunning – and there are a myriad of books and maps out there that will help you with that. But since this book focuses on the "deep sky" objects of stars, star clusters, nebulae and galaxies, I won't be talking about it here. (I *do* recommend you take the time to enjoy the lunar landscape though!)

Image by the author.

If you're going to look at the Moon, I highly recommend you get yourself a lunar filter for your eyepiece, especially if the Moon is between its half phase and being full. The moonlight can be dazzling when magnified through your telescope and the filter will cut out about 85% of that light, making the Moon far easier to observe. Without it, it's like someone shining a flashlight into your eyes. Again, you can easily buy a good filter online and they're not expensive at all.

Let Your 'Scope Adjust to the Air Temperature

Just as your eyes need time to adjust, so does your telescope. Admittedly, this is something I often neglect to do as I'm so keen to get out there that I frequently skip this step. Also, I'm easily distracted and will often find myself getting sidetracked when I should be putting the 'scope outside.

Why is it important? Well, your telescope is probably kept inside and your home is probably kept at a reasonably warm temperature. When you take it outside, the air temperature will be different. It might be warmer or it might be cooler, but until your telescope reaches the same temperature you might find the images are not as sharp as they could be.

If you can handle the wait, take your telescope outside about an hour before it's time to observe. Look at it this way – it can take about ninety minutes for the sky to get dark after sunset anyway, so if you take it outside within the first thirty minutes after the sun goes down and *then* go out after twilight has vanished and it's truly dark, then you should be fine.

Locate the Object with Binoculars First

If you've never observed the object before, using a pair of binoculars can make the search easier and also help you to familiarize yourself with that particular patch of the sky. Many of the objects in this book can be seen with binoculars and a standard pair of 10x50's should do the trick. As always, there are plenty of options available and you can get a good quality pair for about $30 online or through major retailers.

If you're interested in binocular astronomy, there a multitude of books out there but I can highly recommend Philip Harrington's *Touring the Universe through Binoculars* and Gary Seronik's *Binocular Highlights*. I'll often grab my binoculars when I only have a few objects to observe (or too little time) and have included my notes in the descriptive text for each object.

Likewise, many finderscopes will actually magnify the sky a little – often by about 6x. Although not as powerful as binoculars, you should still be able to see almost all the objects in this book within the finder. I've included a graphic depicting the 6x finderscope view beside the text for each object.

Learn to Calculate the Magnification of Your Eyepieces

This is actually very easy but you need to know two important numbers before you begin: the focal length of your telescope and the focal length of the eyepiece you're using.

What's focal length? Well, it's basically the distance that light must travel from entering your telescope (or eyepiece) in order to reach your eye. It's always measured in millimeters and it can always be found on the telescope or eyepiece itself.

Image by the author.

On the telescope, this information can be found on a label that's often affixed to the tube, either on the side or close to the focuser where the eyepiece is inserted. It'll probably have the make and model of the telescope and the focal length may be represented with the letters FL.

On an eyepiece, it's printed at the top, almost always on the side and sometimes around the lens of the eyepiece itself.

To calculate the magnification of the eyepiece, you simply divide the telescope's focal length by the focal length of the eyepiece. For example, most telescopes come with a 10mm eyepiece (meaning it has a focal length of 10mm.) My Celestron 130 SLT GoTo telescope has a focal length of 650mm so a 10mm eyepiece will give me a magnification of 65x.

Get Yourself a Range of Eyepieces

Obviously then, it's useful to have a small number of eyepieces as this will give you a good range of magnifications to play around with. Most telescopes seem to come with a fairly large eyepiece (for example, a 20mm or 25mm) that will give a low powered view and also a smaller eyepiece (typically a 10mm or a 6mm) that will give a higher powered view.

Personally, I wouldn't go any lower than 6mm because the lens and field of view typically decreases in relation to the focal length. This means you have to put your eye closer to the eyepiece and, in my experience, this actually makes observing a little uncomfortable. (There's a way around that – it's called a barlow and I'll talk about that in a moment.)

Likewise, unless my telescope had a high focal length (say, over 900mm) I wouldn't go over 30mm. (You can, for example, buy 50mm eyepieces but they tend to be a little pricey.)

Some of the eyepieces I own. From left to right, 25mm, 20mm, 15mm, 10mm and a 6mm. Image by the author.

I'd recommend a 25mm and then adding eyepieces in 5mm increments. I therefore own a 25mm, 20mm, 15mm and 10mm along with a 6mm and a 9mm. (Once you get to 10mm, the sizes tend to decrease in smaller increments.)

You don't have to buy these individually either. Both Orion and Celestron produce eyepiece kits for beginners that will give you a good range to start off with. My only word of caution is to make sure you get the right barrel size for your telescope. The vast majority of telescopes have a focuser that will fit eyepieces with a barrel 1.25 inches wide, but there are some that will take slightly different sizes. If you're not sure, check your telescope's manual or contact the manufacturer before buying.

But why do you need a range of magnification anyway? Surely a high and a low powered eyepiece is enough? Potentially, yes, but you'll soon learn that when it comes to observing deep sky objects, there comes a point when you magnify an object *too* much and it loses its aesthetic appeal. A star cluster is prettier when you can see the whole cluster and it's set against a backdrop of faint stars. Multiple stars (especially those of differing colors) are more attractive when they appear close together, rather than highly magnified and far apart.

Think you only need the two that came with your 'scope? Observe the deep sky objects in this book and then see how you feel!

Know the Maximum Useful Magnification of Your Telescope

Unfortunately, as I mentioned in the Introduction, your telescope is probably not going to produce a good quality image at 1,000x. In fact, I'd be impressed if you can even squeeze 1,000x out of your 'scope. There are a number of reasons for this, but suffice it to say, every telescope has its maximum useful magnification and it's important to know this before you start observing (or start buying eyepieces that go beyond this limit.)

Quite simply, the theoretical maximum magnification is 50x the aperture of your telescope in inches or about twice the aperture in millimeters. For example, my Orion SkyQuest XT 4.5 Dobsonian had an aperture of 4½ inches so the theoretical maximum was 225x. (4½ inches in millimeters is slightly more than 114, so that would be 228x.)

However, I've found that this isn't always the case. Realistically, I've found the views can be unreliable beyond half this amount. Sometimes they're better, but more often they're not. (For example, the only time I've seen the Great Red Spot on Jupiter with my Celestron 130 SLT is when I had it right at the maximum of 260x. However, I don't tend to notice much improvement in many faint and fuzzy deep sky objects.)

Therefore I'd go for 25x the aperture of your telescope in inches or – more conveniently – the aperture of your telescope in millimeters. So I usually stick to 130x or less on my Celestron 130 SLT and have never been disappointed as a result.

Invest in a Barlow Lens

I've recommended that you buy yourself a range of eyepieces and although you can buy these for a reasonable price, some can be quite expensive. Also, the smaller the focal length of your eyepiece, the smaller your field of view, which can make it hard to observe.

The solution? A 2x barlow lens. This is basically a short tube that slots into your telescope's focuser, where the eyepiece normally goes. You then pop your eyepiece into the barlow and *voila!* Your magnification is instantly doubled. And the best part is your field of view remains the same.

For example, a 20mm eyepiece in my Celestron 130 SLT will give me a magnification of about 32½x with a nice, wide, comfortable field of view. But if I'm using a 2x barlow, I double up to 65x and still be comfortable with the same field of view.

A 2x barlow. Image by the author.

Barlows are very easily obtained, are reasonably priced and the advantages are obvious. Better magnification. A more enjoyable observing experience. Potentially few eyepieces needed so you won't have to carry as much equipment around. It's easier on the eyes *and* the wallet.

Buy a Toolbox for Your Equipment

Yes, a toolbox. Or a fishing tackle box. One of those boxes you buy in hardware stores to keep small tools, screws and odds and ends does the job quite nicely.

Mine has a lower and an upper compartment; in the lower I keep the things I might only need to use once or a few times in a session: my GoTo power cable, a pocket night sky guide and a mini planisphere.

In the upper compartment, I keep the items I'll use frequently, such as my eyepieces, my barlow, my red flashlight and my filters.

And all I have to do is carry it outside. Trust me, it's invaluable.

Image by the author.

Start with a Lower Power Eyepiece and Work Your Way Up

When I'm observing, I always start at a low magnification and then increase it. For starters, it's easier to find your target and many star clusters are best observed at low magnification anyway. If the object is easily seen, then I'll gradually increase the power to see how the view changes.

If, however, the target isn't seen, then I'll check to make sure I'm looking in the right place and then up the power to something closer to 100x. This is also true of multiple stars that can't be split at low magnification. For example, I may only see a single, bright star at 35x so I'll up the power to around 100x and see if my luck improves. If the star's split, I'll start decreasing the magnification to find the point where it becomes a single star again. Quite frequently, I've gotten down to 35x again and still been able to split the star – but only because I now know where to look for the companion.

Focus on the Stars

If you're staring at something faint and fuzzy (for example, a globular star cluster) you'll often find it difficult to properly focus the view. Instead of trying to focus on the object, focus on one of the nearby stars instead. And if you don't see any in the same field of view, try moving the 'scope slightly so that some are visible. It's far easier to focus on a single point of light than something grey, misty and ill-defined!

The Colors of Multiple Stars

I've found that star colors can be very subjective and may even seem to vary from night to night. There are a number of reasons for this. Unless the colors are quite strong (such as the famous double star Albireo) what you see might vary depending upon the sky conditions, the equipment you're using, the height of the star in the sky and, frankly simply, your own eyesight.

So if I've described a blue-white star and you're seeing just white – or maybe even pale gold or yellow – then there's nothing wrong. It just means that, like almost everything in life, you're having a slightly different experience. (If you read the observations of others, you'll often find a wide range of colors being described.)

If you're having difficulty seeing any color at all, try de-focusing the view. As the star loses clarity, it will appear larger, spreading the light over a wider area in your field of view and sometimes making the color easier to see.

Use Averted Vision

What's averted vision? Basically, it's observing the object out of the corner of your eye. Without getting too technical, your peripheral vision is more sensitive to light (and movement) than your direct vision. It's thought this enabled our ancestors to detect potential nearby predators before they became a more immediate threat.

If you can't see an object, try using averted vision. If that doesn't work, try tapping the telescope tube gently to make the view move as averted vision is also good for detecting movement.

Once you've found it, looking at the object with averted vision can reveal detail not visible with direct observation. However, there's a trade-off.

From personal experience, it seems that while averted vision may help you to see a faint object, it's usually lacking in color, whereas direct vision may show shades of green or blue. The center of your eye is better suited for providing a good, all-round color view while it seems the edge of your vision is best suited for detection and motion.

Keep an Astronomical Journal

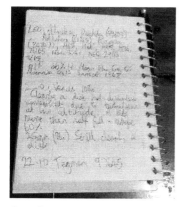

One of my journals, complete with chicken-scratch handwriting. Image by the author.

As a kid I used to keep one and wrote an entry for every object I observed on every night. Well, almost. I admit to skipping it sometimes. There's a couple of good reasons for this. Firstly, it's a great way to see keep track of which objects you've seen and to feel a real sense of accomplishment as your tally increases.

Secondly, I use it to plan my session. Before I go outside, I'll make a list of all the objects I want to observe, along with any additional information I'll need – for example, the object's catalogue number or co-ordinates. As I'm going outside, I'll also make a note of the weather, sunset time and Moon information such as its phase, illumination and rise or set time.

Lastly, it's simply a good reference tool. It's a rare object that only gets observed once and you'll return to many of your favorite objects over and over again. However, I don't re-read my notes before I observe that object again. Why? Because it can actually prejudice your viewing experience and influence your expectations.

I've also found it sometimes spoils the experience a little. There's nothing better than observing an object for the first time and discovering something new and when you're starting out there's literally hundreds of objects to discover. If you don't review your notes beforehand, it's easy to forget what you've seen and then you go to the eyepiece and rediscover that object all over again. (There's an exception to this rule. Sometimes I'll observe an object but make a separate note to re-observe it later, looking for a specific detail.)

Similarly, unless you're having difficulty observing an object, try to avoid the notes of others beforehand. I know this sounds contrary to the concept of this book, but I'm hoping the text will give you an idea of what to expect while encouraging you to have your own personal experience. And that's really the point of astronomy. It's a personal experience for everyone. Reading someone else's opinion of an object beforehand can color that experience and, I feel, you lose something as a result.

Compare your notes after you're done for the night, not before. That way you can find out if that star was truly multiple or if you were just imagining it. Or you'll read about a detail you missed and you'll want to go back and observe that object again.

Go, observe, keep a journal and compare notes later. You'll enjoy it far more that way.

Signposts to the Stars

Ursa Major

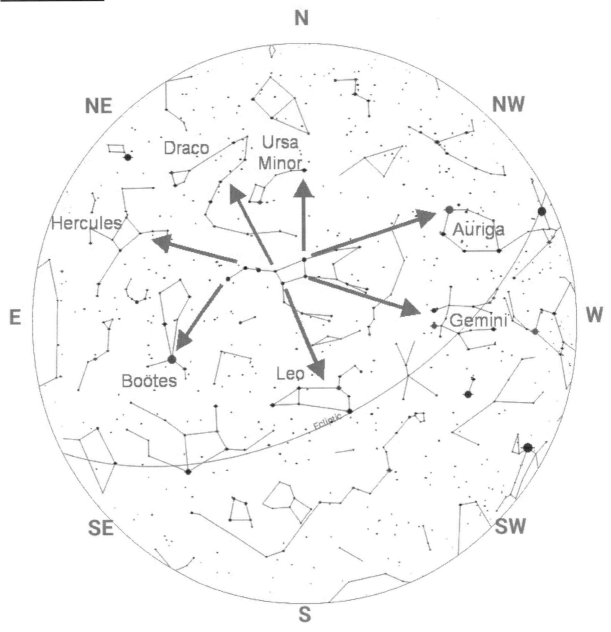

You can use the brightest stars in Ursa Major, commonly known as the Big Dipper (or the Plough in the United Kingdom) to find a number of other constellations. The table below lists the objects to be seen in those constellations and the map above is similar to star chart #12.

Object	Page
Messier 37 (Auriga)	90
Messier 35 (Gemini)	91
Castor (Gemini)	92
Polaris (Ursa Minor)	82
Regulus (Leo)	101

Object	Page
Adhafera (Leo)	102
Algieba (Leo)	103
Denebola (Leo)	104
Arcturus (Boötes)	108

Object	Page
Delta Boötis (Boötes)	109
Keystone Cluster (Hercules)	117
Rho Herculis (Hercules)	118
Kuma (Draco)	119

Orion

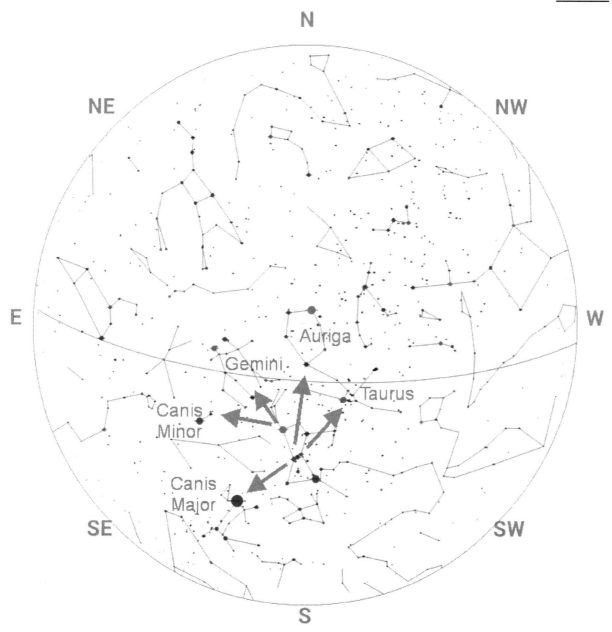

You can use Orion to find a number of other constellations. The table below lists the objects to be seen in those constellations and the map above is similar to star chart #6.

Object	Page
Pleiades (Taurus)	83
Crab Nebula (Taurus)	84
Messier 37 (Auriga)	90

Object	Page
Messier 35 (Gemini)	91
Castor (Gemini)	92
Procyon (Canis Minor)	96

Object	Page
Sirius (Canis Major)	94
Messier 41 (Canis Major)	95

Star Charts & Observing Lists

Star Chart Tables

If observing during daylight savings time, first deduct one hour and then refer to the corresponding chart number. For example, for 10pm daylight savings time in early August, use chart 18.

	6pm	7pm	8pm	9pm	10pm	11pm	12am	1am	2am
Early January	1	2	3	4	5	6	7	8	9
Late January	2	3	4	5	6	7	8	9	10
Early February	3	4	5	6	7	8	9	10	11
Late February	4	5	6	7	8	9	10	11	12
Early March	5	6	7	8	9	10	11	12	13
Late March	6	7	8	9	10	11	12	13	14
Early April	7	8	9	10	11	12	13	14	15
Late April	8	9	10	11	12	13	14	15	16
Early May	9	10	11	12	13	14	15	16	17
Late May	10	11	12	13	14	15	16	17	18
Early June	11	12	13	14	15	16	17	18	19
Late June	12	13	14	15	16	17	18	19	20
Early July	13	14	15	16	17	18	19	20	21
Late July	14	15	16	17	18	19	20	21	22
Early August	15	16	17	18	19	20	21	22	23
Late August	16	17	18	19	20	21	22	23	24
Early September	17	18	19	20	21	22	23	24	1
Late September	18	19	20	21	22	23	24	1	2
Early October	19	20	21	22	23	24	1	2	3
Late October	20	21	22	23	24	1	2	3	4
Early November	21	22	23	24	1	2	3	4	5
Late November	22	23	24	1	2	3	4	5	6
Early December	23	24	1	2	3	4	5	6	7
Late December	24	1	2	3	4	5	6	7	8

Chart 1

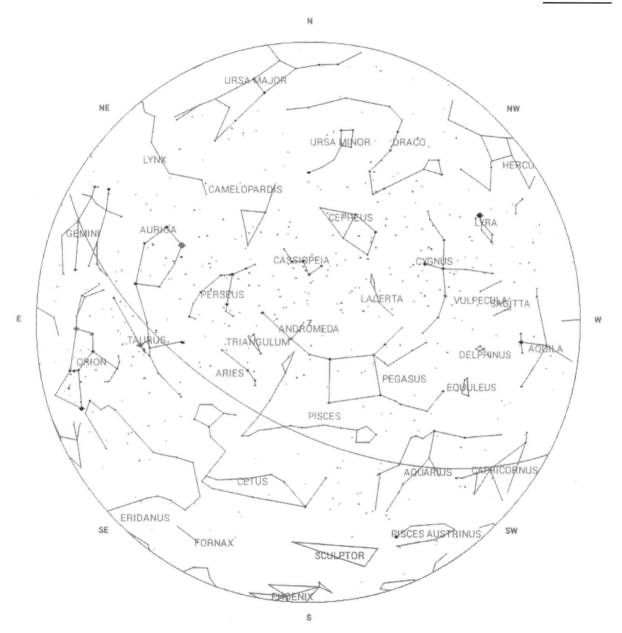

NB: Not all visible objects may be listed. Objects are listed based upon their visibility, from west to east.

Object	Page
Kuma	119
Double Double	121
Sheliak	122
Albireo	127
Messier 29	128
Zeta Sagittae	129
Gamma Delphini	132
Messier 15	133

Object	Page
Pi Andromedae	70
Andromeda Galaxy	71
NGC 752	72
Almach	73
Achird	74
Owl Cluster	75
Messier 103	76
NGC 663	77

Object	Page
Double Cluster	80
Messier 34	81
Mesarthim	78
Lambda Arietis	79
Pleiades	83
Messier 37	90
Messier 35	91
Meissa	86

Chart 2

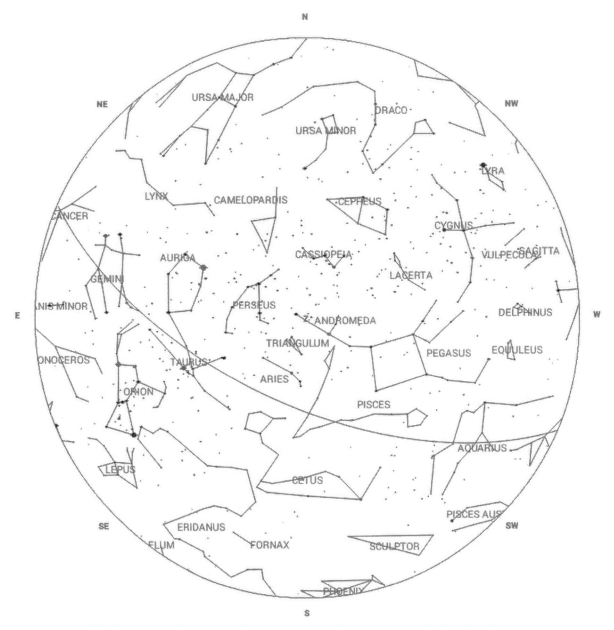

NB: Not all visible objects may be listed. Objects are listed based upon their visibility, from west to east.

Object	Page
Messier 29	128
Messier 15	133
Pi Andromedae	70
Andromeda Galaxy	71
NGC 752	72
Almach	73
Achird	74
Owl Cluster	75

Object	Page
Messier 103	76
NGC 663	77
Double Cluster	80
Messier 34	81
Mesarthim	78
Lambda Arietis	79
Pleiades	83
Crab Nebula	84

Object	Page
Messier 37	90
Messier 35	91
Castor	92
Meissa	86
Mintaka	85
Sigma Orionis	88
Orion Nebula	87
Polaris	82

Chart 3

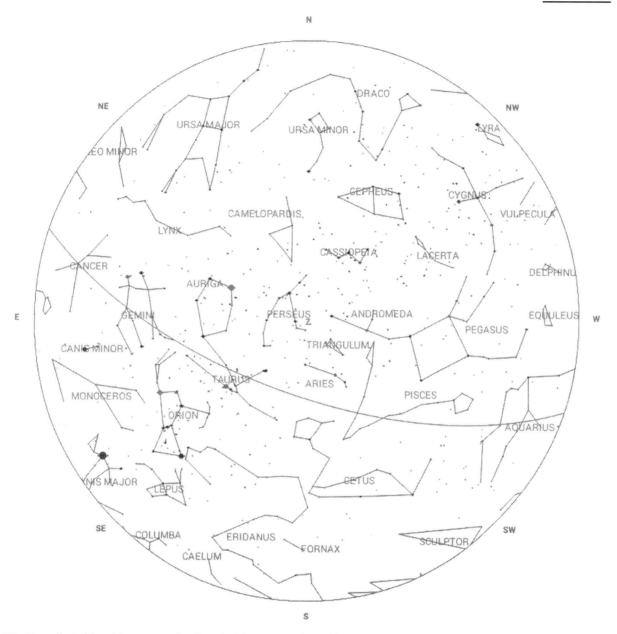

NB: Not all visible objects may be listed. Objects are listed based upon their visibility, from west to east.

Object	Page
Pi Andromedae	70
Andromeda Galaxy	71
NGC 752	72
Almach	73
Achird	74
Owl Cluster	75
Messier 103	76
NGC 663	77

Object	Page
Double Cluster	80
Messier 34	81
Mesarthim	78
Lambda Arietis	79
Pleiades	83
Crab Nebula	84
Messier 37	90
Messier 35	91

Object	Page
Castor	92
Meissa	86
Mintaka	85
Sigma Orionis	88
Orion Nebula	87
Gamma Leporis	89
Beta Monocerotis	93
Polaris	82

Chart 4

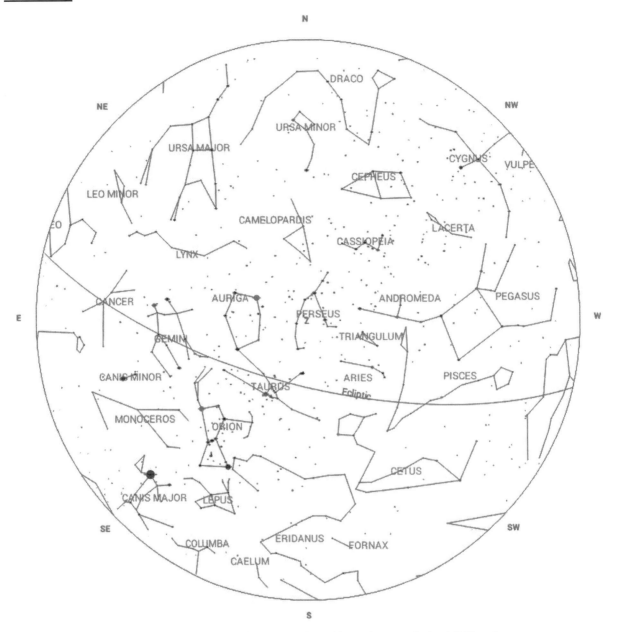

NB: Not all visible objects may be listed. Objects are listed based upon their visibility, from west to east.

Object	Page
Pi Andromedae	70
Andromeda Galaxy	71
NGC 752	72
Almach	73
Achird	74
Owl Cluster	75
Messier 103	76
NGC 663	77

Object	Page
Double Cluster	80
Messier 34	81
Mesarthim	78
Lambda Arietis	79
Pleiades	83
Crab Nebula	84
Messier 37	90
Messier 35	91

Object	Page
Castor	92
Meissa	86
Mintaka	85
Sigma Orionis	88
Orion Nebula	87
Gamma Leporis	89
Beta Monocerotis	93
Procyon	96

Chart 5

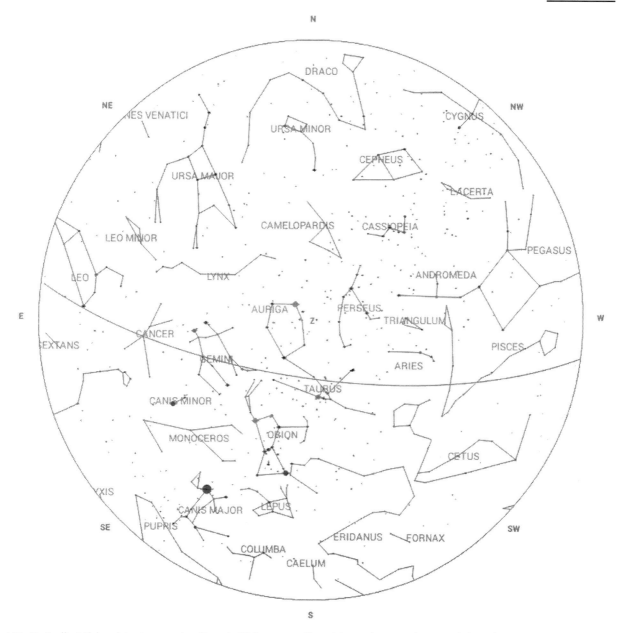

NB: Not all visible objects may be listed. Objects are listed based upon their visibility, from west to east.

Object	Page
Pi Andromedae	70
Andromeda Galaxy	71
NGC 752	72
Almach	73
Achird	74
Owl Cluster	75
Messier 103	76
NGC 663	77

Object	Page
Double Cluster	80
Messier 34	81
Mesarthim	78
Lambda Arietis	79
Pleiades	83
Crab Nebula	84
Messier 37	90
Messier 35	91

Object	Page
Castor	92
Meissa	86
Mintaka	85
Sigma Orionis	88
Orion Nebula	87
Gamma Leporis	89
Sirius	94
Beta Monocerotis	93

Chart 6

NB: Not all visible objects may be listed. Objects are listed based upon their visibility, from west to east.

Object	Page
Andromeda Galaxy	71
NGC 752	72
Almach	73
Achird	74
Owl Cluster	75
Messier 103	76
NGC 663	77
Double Cluster	80

Object	Page
Messier 34	81
Mesarthim	78
Lambda Arietis	79
Pleiades	83
Crab Nebula	84
Messier 37	90
Messier 35	91
Castor	92

Object	Page
Meissa	86
Mintaka	85
Sigma Orionis	88
Orion Nebula	87
Gamma Leporis	89
Sirius	94
Messier 41	95
Beta Monocerotis	93

Chart 7

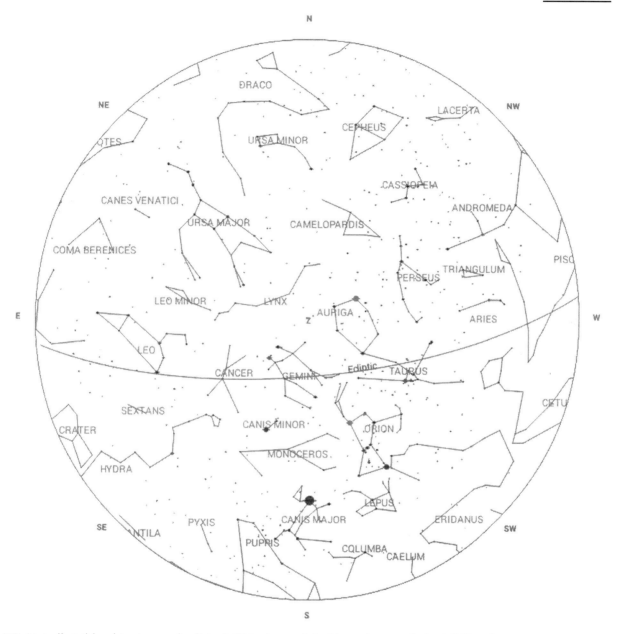

NB: Not all visible objects may be listed. Objects are listed based upon their visibility, from west to east.

Object	Page
Achird	74
Owl Cluster	75
Messier 103	76
NGC 663	77
Double Cluster	80
Messier 34	81
Pleiades	83
Crab Nebula	84

Object	Page
Messier 37	90
Messier 35	91
Castor	92
Meissa	86
Mintaka	85
Sigma Orionis	88
Orion Nebula	87
Gamma Leporis	89

Object	Page
Sirius	94
Messier 41	95
Messier 93	97
Beta Monocerotis	93
Procyon	96
Iota Cancri	99
Praesepe	98
Messier 67	100

Chart 8

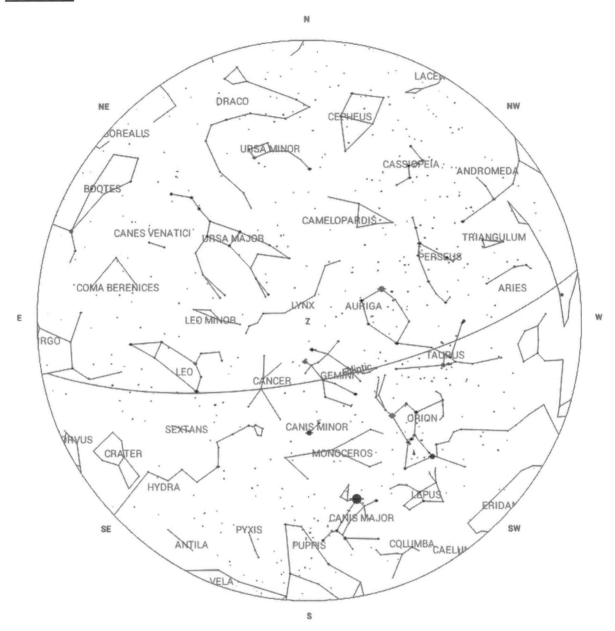

NB: Not all visible objects may be listed. Objects are listed based upon their visibility, from west to east.

Object	Page
Double Cluster	80
Messier 34	81
Pleiades	83
Crab Nebula	84
Messier 37	90
Messier 35	91
Castor	92
Meissa	86

Object	Page
Mintaka	85
Sigma Orionis	88
Orion Nebula	87
Gamma Leporis	89
Sirius	94
Messier 41	95
Messier 93	97
Beta Monocerotis	93

Object	Page
Procyon	96
Iota Cancri	99
Praesepe	98
Messier 67	100
Regulus	101
Algieba	103
Adhafera	102
Denebola	104

Chart 9

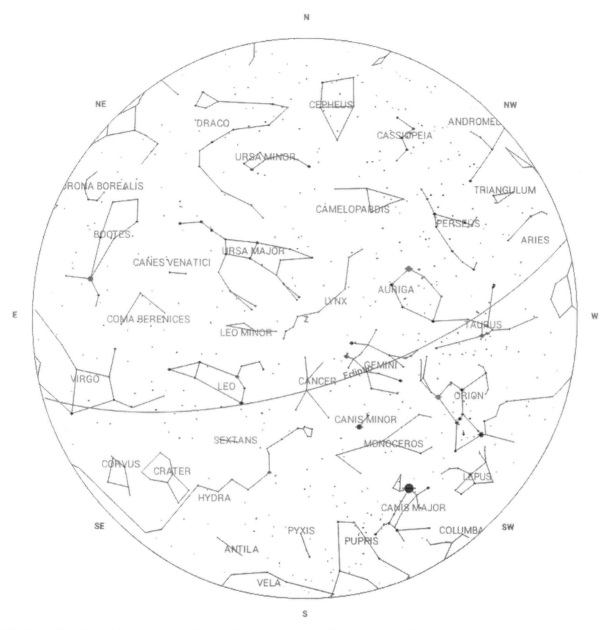

NB: Not all visible objects may be listed. Objects are listed based upon their visibility, from west to east.

Object	Page
Pleiades	83
Crab Nebula	84
Messier 37	90
Messier 35	91
Castor	92
Meissa	86
Mintaka	85
Sigma Orionis	88

Object	Page
Orion Nebula	87
Sirius	94
Messier 41	95
Messier 93	97
Beta Monocerotis	93
Procyon	96
Iota Cancri	99
Praesepe	98

Object	Page
Messier 67	100
Regulus	101
Algieba	103
Adhafera	102
Denebola	104
Mizar & Alcor	107
Cor Caroli	106
Arcturus	108

Chart 10

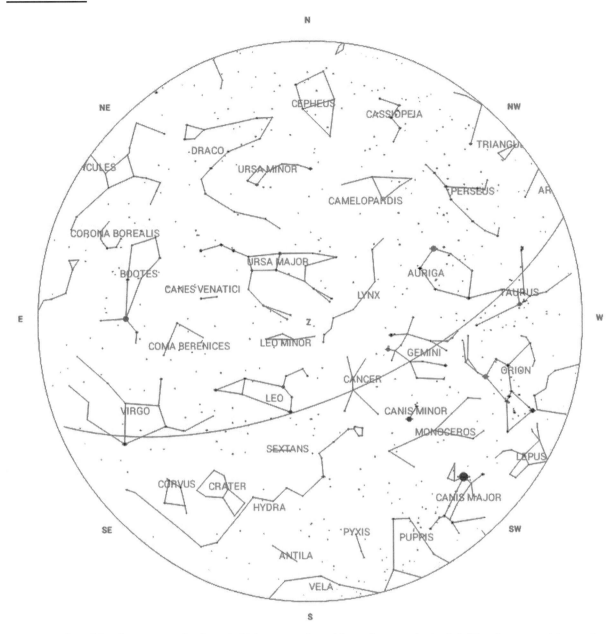

NB: Not all visible objects may be listed. Objects are listed based upon their visibility, from west to east.

Object	Page
Messier 37	90
Messier 35	91
Castor	92
Meissa	86
Sirius	94
Messier 41	95
Messier 93	97

Object	Page
Beta Monocerotis	93
Procyon	96
Iota Cancri	99
Praesepe	98
Messier 67	100
Regulus	101
Algieba	103

Object	Page
Adhafera	102
Denebola	104
Polaris	82
Mizar & Alcor	107
Cor Caroli	106
Arcturus	108
Delta Boötis	109

Chart 11

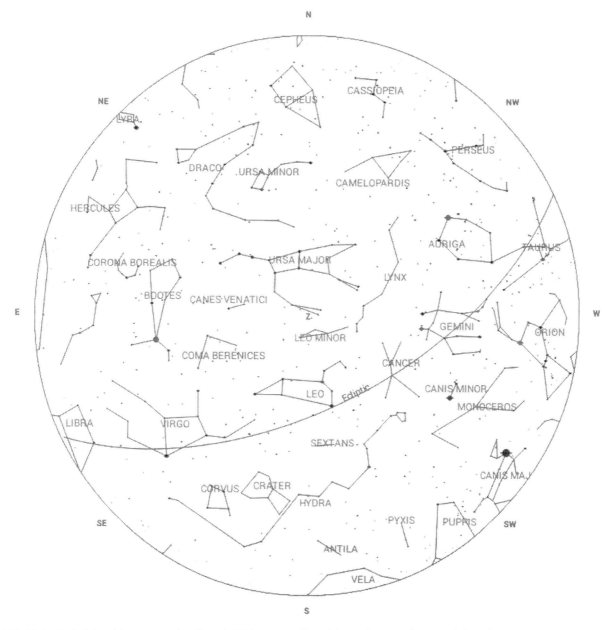

NB: Not all visible objects may be listed. Objects are listed based upon their visibility, from west to east.

Object	Page
Messier 37	90
Messier 35	91
Castor	92
Beta Monocerotis	93
Procyon	96
Iota Cancri	99
Praesepe	98

Object	Page
Messier 67	100
Regulus	101
Algieba	103
Adhafera	102
Denebola	104
Algorab	105

Object	Page
Polaris	82
Kuma	119
Mizar & Alcor	107
Cor Caroli	106
Arcturus	108
Delta Boötis	109

Chart 12

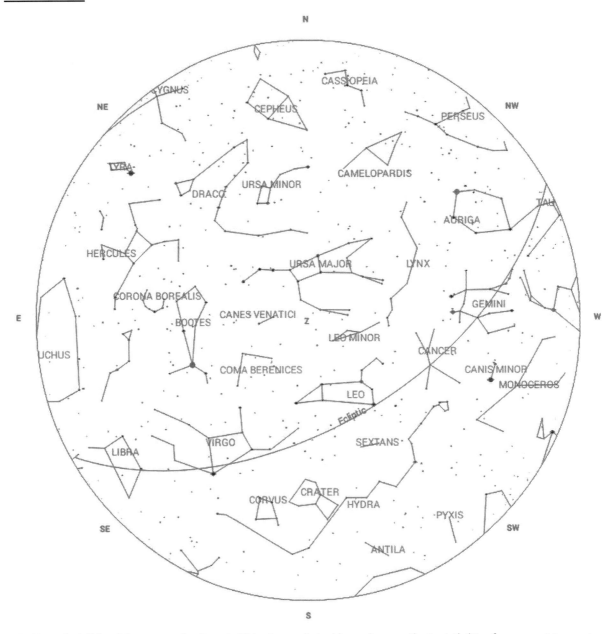

NB: Not all visible objects may be listed. Objects are listed based upon their visibility, from west to east.

Object	Page
Castor	92
Procyon	96
Iota Cancri	99
Praesepe	98
Messier 67	100
Regulus	101

Object	Page
Algieba	103
Adhafera	102
Denebola	104
Algorab	105
Polaris	82
Mizar & Alcor	107

Object	Page
Cor Caroli	106
Arcturus	108
Delta Boötis	109
Kuma	119
Keystone Cluster	117
Rho Herculis	118

Chart 13

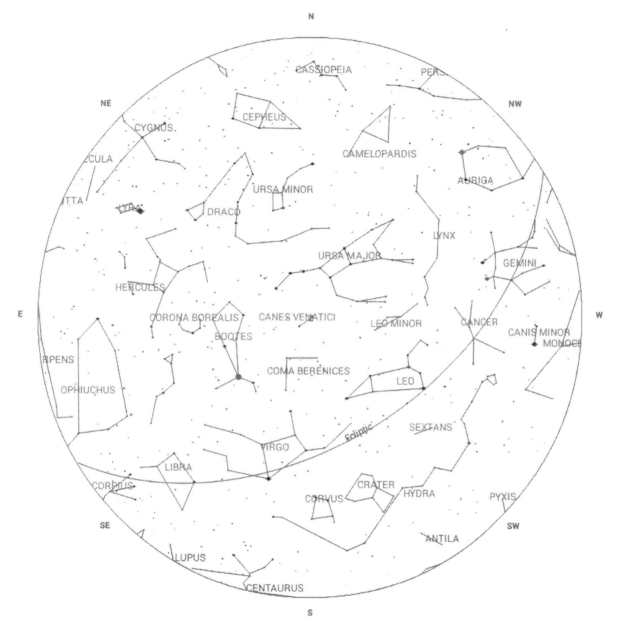

NB: Not all visible objects may be listed. Objects are listed based upon their visibility, from west to east.

Object	Page
Castor	92
Iota Cancri	99
Praesepe	98
Messier 67	100
Regulus	101
Algieba	103

Object	Page
Adhafera	102
Denebola	104
Algorab	105
Polaris	82
Mizar & Alcor	107
Cor Caroli	106

Object	Page
Arcturus	108
Delta Boötis	109
Keystone Cluster	117
Rho Herculis	118
Kuma	119
Zuben Elgenubi	110

Chart 14

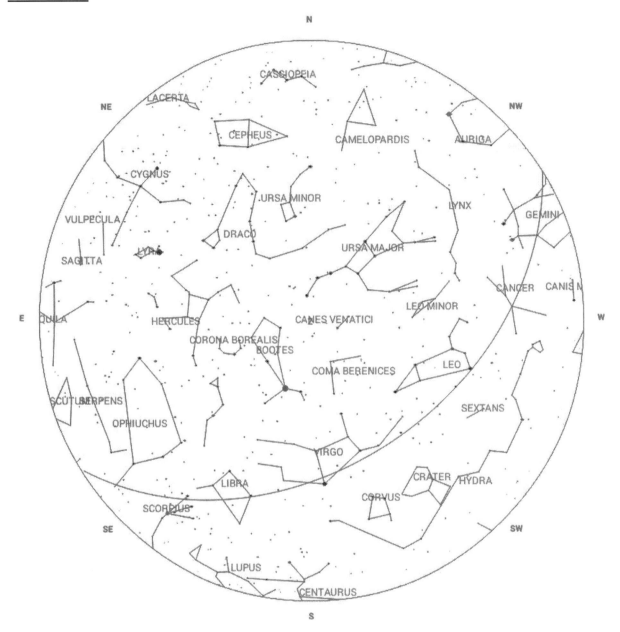

NB: Not all visible objects may be listed. Objects are listed based upon their visibility, from west to east.

Object	Page
Iota Cancri	99
Praesepe	98
Messier 67	100
Regulus	101
Algieba	103
Adhafera	102
Denebola	104

Object	Page
Algorab	105
Polaris	82
Mizar & Alcor	107
Cor Caroli	106
Arcturus	108
Delta Boötis	109

Object	Page
Zuben Elgenubi	110
Keystone Cluster	117
Rho Herculis	118
Kuma	119
Double Double	121
Sheliak	122

Chart 15

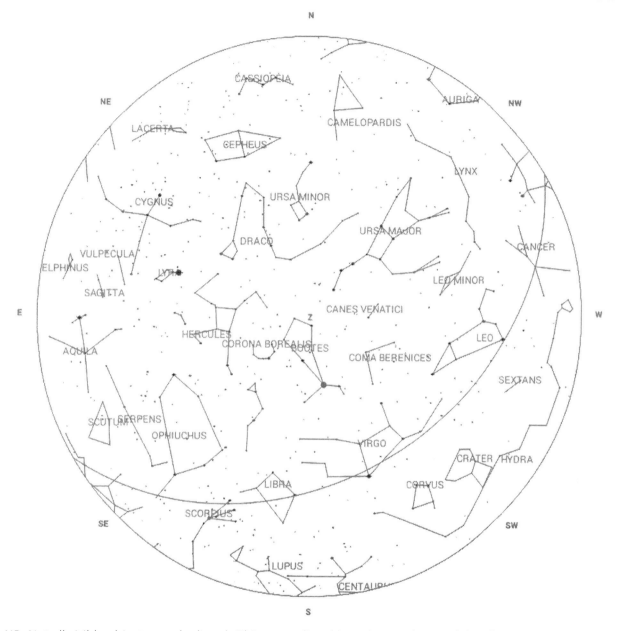

NB: Not all visible objects may be listed. Objects are listed based upon their visibility, from west to east.

Object	Page
Regulus	101
Algieba	103
Adhafera	102
Denebola	104
Algorab	105
Polaris	82
Mizar & Alcor	107

Object	Page
Cor Caroli	106
Arcturus	108
Delta Boötis	109
Zuben Elgenubi	110
Graffias	111
Jabbah	112
Keystone Cluster	117

Object	Page
Rho Herculis	118
Kuma	119
Double Double	121
Sheliak	122
Ring Nebula	123
Albireo	127
Messier 29	128

Chart 16

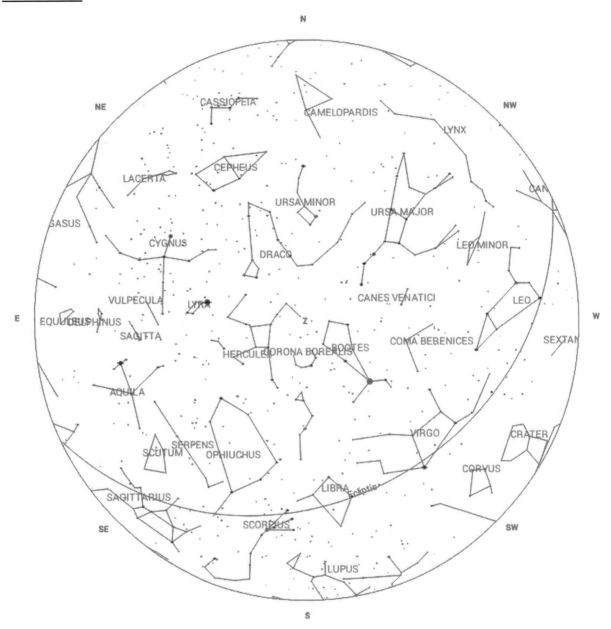

NB: Not all visible objects may be listed. Objects are listed based upon their visibility, from west to east.

Object	Page
Denebola	104
Polaris	82
Mizar & Alcor	107
Cor Caroli	106
Arcturus	108
Delta Boötis	109
Zuben Elgenubi	110
Graffias	111

Object	Page
Jabbah	112
Messier 80	113
Messier 4	114
Keystone Cluster	117
Rho Herculis	118
Kuma	119
Double Double	121
Sheliak	122

Object	Page
Ring Nebula	123
Albireo	127
Messier 29	128
Dumbbell Nebula	126
Coathanger Cluster	125
Zeta Sagittae	129
Messier 71	130
Wild Duck Cluster	124

Chart 17

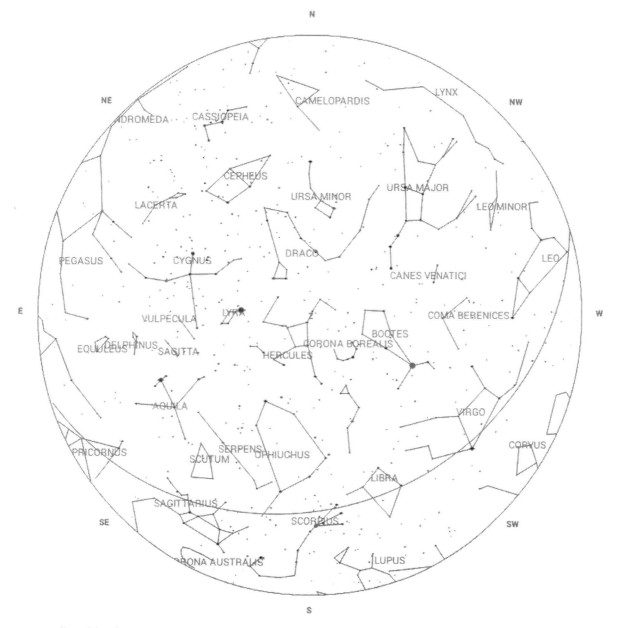

NB: Not all visible objects may be listed. Objects are listed based upon their visibility, from west to east.

Object	Page
Polaris	82
Mizar & Alcor	107
Cor Caroli	106
Arcturus	108
Delta Boötis	109
Zuben Elgenubi	110
Graffias	111
Jabbah	112

Object	Page
Messier 80	113
Messier 4	114
Keystone Cluster	117
Rho Herculis	118
Kuma	119
Double Double	121
Sheliak	122
Ring Nebula	123

Object	Page
Albireo	127
Messier 29	128
Dumbbell Nebula	126
Coathanger Cluster	125
Zeta Sagittae	129
Messier 71	130
Gamma Delphini	132
Wild Duck Cluster	124

Chart 18

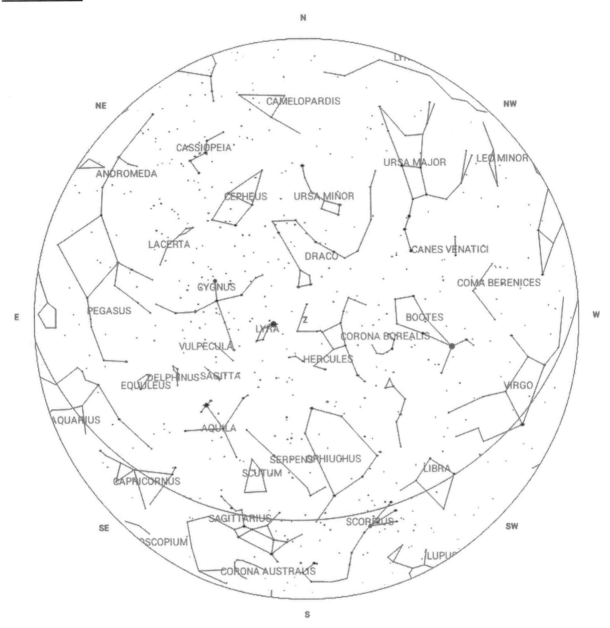

NB: Not all visible objects may be listed. Objects are listed based upon their visibility, from west to east.

Object	Page
Arcturus	108
Delta Boötis	109
Zuben Elgenubi	110
Graffias	111
Jabbah	112
Messier 80	113
Messier 4	114
Keystone Cluster	117

Object	Page
Rho Herculis	118
Kuma	119
Double Double	121
Sheliak	122
Ring Nebula	123
Albireo	127
Messier 29	128
Dumbbell Nebula	126

Object	Page
Coathanger Cluster	125
Zeta Sagittae	129
Messier 71	130
Gamma Delphini	132
Wild Duck Cluster	124
Messier 22	120
Butterfly Cluster	115
Messier 7	116

Chart 19

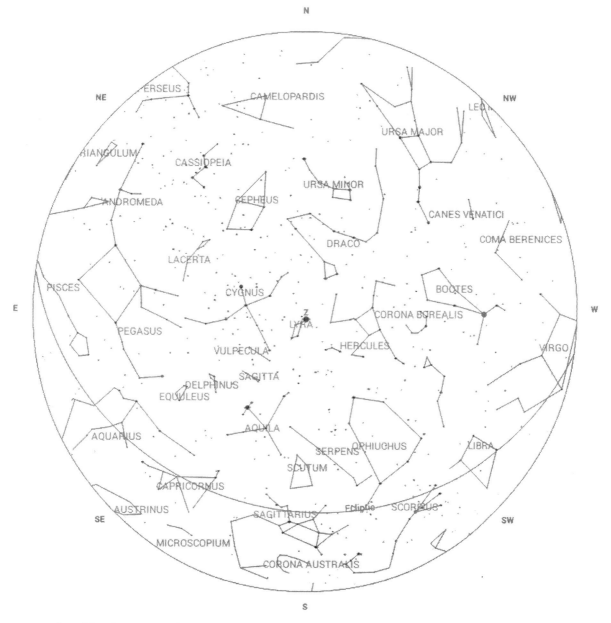

NB: Not all visible objects may be listed. Objects are listed based upon their visibility, from west to east.

Object	Page
Arcturus	108
Delta Boötis	109
Graffias	111
Jabbah	112
Keystone Cluster	117
Rho Herculis	118
Kuma	119
Double Double	121

Object	Page
Sheliak	122
Ring Nebula	123
Albireo	127
Messier 29	128
Dumbbell Nebula	126
Coathanger Cluster	125
Zeta Sagittae	129
Messier 71	130

Object	Page
Gamma Delphini	132
Messier 15	133
Al Giedi	131
Wild Duck Cluster	124
Messier 22	120
Butterfly Cluster	115
Messier 7	116

Chart 20

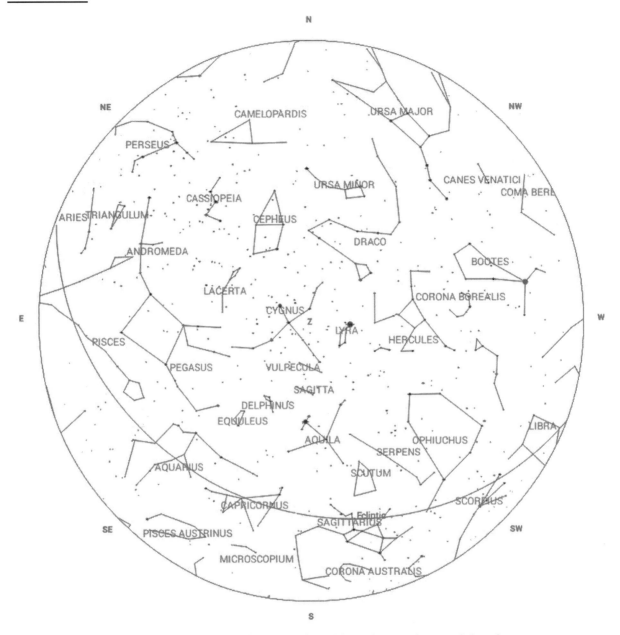

NB: Not all visible objects may be listed. Objects are listed based upon their visibility, from west to east.

Object	Page
Delta Boötis	109
Keystone Cluster	117
Rho Herculis	118
Kuma	119
Double Double	121
Sheliak	122
Ring Nebula	123
Albireo	127

Object	Page
Messier 29	128
Dumbbell Nebula	126
Coathanger Cluster	125
Zeta Sagittae	129
Messier 71	130
Gamma Delphini	132
Messier 15	133
Al Giedi	131

Object	Page
Wild Duck Cluster	124
Messier 22	120
Pi Andromedae	70
Achird	74
Owl Cluster	75
Messier 103	76
NGC 663	77

Chart 21

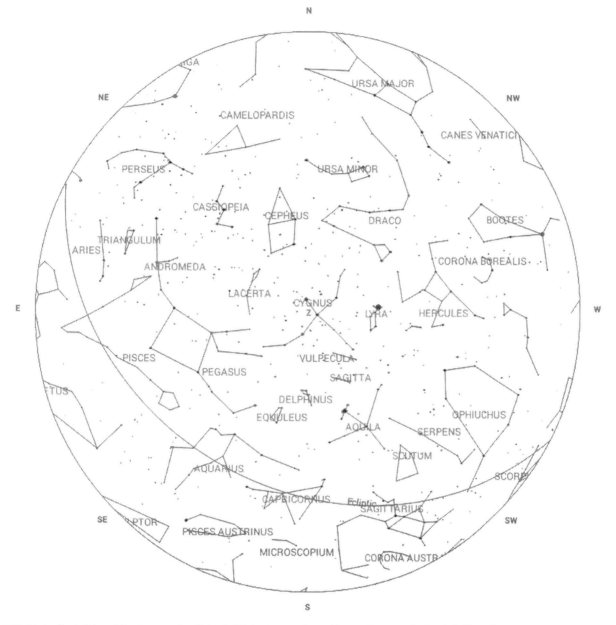

NB: Not all visible objects may be listed. Objects are listed based upon their visibility, from west to east.

Object	Page
Keystone Cluster	117
Rho Herculis	118
Kuma	119
Double Double	121
Sheliak	122
Ring Nebula	123
Albireo	127
Messier 29	128

Object	Page
Dumbbell Nebula	126
Coathanger Cluster	125
Zeta Sagittae	129
Messier 71	130
Gamma Delphini	132
Messier 15	133
Al Giedi	131
Wild Duck Cluster	124

Object	Page
Pi Andromedae	70
Andromeda Galaxy	71
Almach	73
Achird	74
Owl Cluster	75
Messier 103	76
NGC 663	77

Chart 22

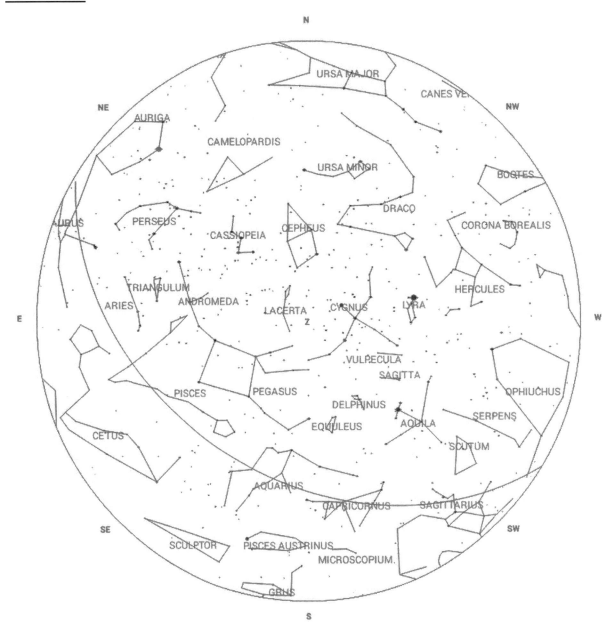

NB: Not all visible objects may be listed. Objects are listed based upon their visibility, from west to east.

Object	Page
Keystone Cluster	117
Rho Herculis	118
Kuma	119
Double Double	121
Sheliak	122
Ring Nebula	123
Albireo	127
Messier 29	128

Object	Page
Dumbbell Nebula	126
Coathanger Cluster	125
Zeta Sagittae	129
Messier 71	130
Gamma Delphini	132
Messier 15	133
Al Giedi	131
Wild Duck Cluster	124

Object	Page
Pi Andromedae	70
Andromeda Galaxy	71
Almach	73
Achird	74
Owl Cluster	75
Messier 103	76
NGC 663	77
Double Cluster	80

Chart 23

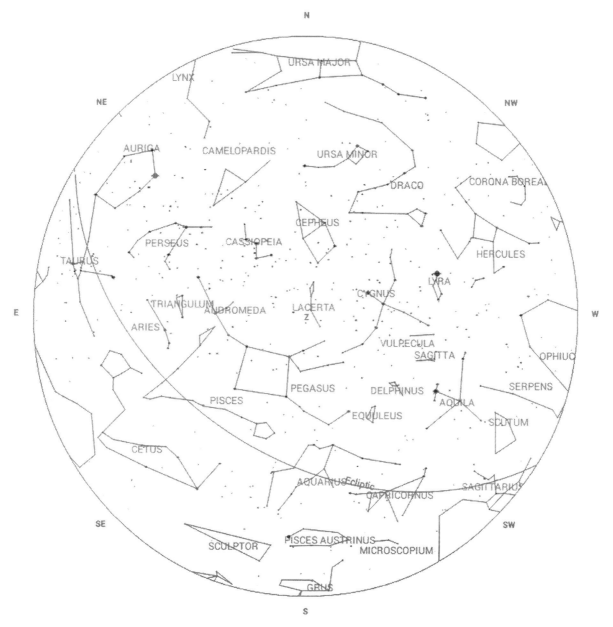

NB: Not all visible objects may be listed. Objects are listed based upon their visibility, from west to east.

Object	Page
Kuma	119
Double Double	121
Sheliak	122
Ring Nebula	123
Albireo	127
Messier 29	128
Dumbbell Nebula	126
Coathanger Cluster	125

Object	Page
Zeta Sagittae	129
Messier 71	130
Gamma Delphini	132
Messier 15	133
Pi Andromedae	70
Andromeda Galaxy	71
Almach	73
Achird	74

Object	Page
Owl Cluster	75
Messier 103	76
NGC 663	77
Double Cluster	80
Messier 34	81
NGC 752	72
Lambda Arietis	79
Mesarthim	78

Chart 24

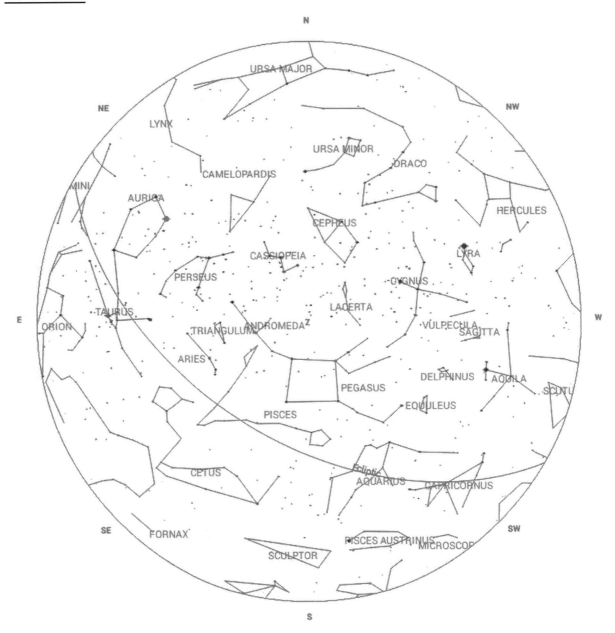

NB: Not all visible objects may be listed. Objects are listed based upon their visibility, from west to east.

Object	Page
Double Double	121
Sheliak	122
Ring Nebula	123
Albireo	127
Messier 29	128
Dumbbell Nebula	126
Coathanger Cluster	125
Zeta Sagittae	129

Object	Page
Messier 71	130
Gamma Delphini	132
Messier 15	133
Pi Andromedae	70
Andromeda Galaxy	71
Almach	73
Achird	74
Owl Cluster	75

Object	Page
Messier 103	76
NGC 663	77
Double Cluster	80
Messier 34	81
NGC 752	72
Lambda Arietis	79
Mesarthim	78
Pleiades	83

The Things You Can See

About the Deep Sky Objects

There are basically five types of objects you can observe with your telescope:

1. Solar System Objects
2. Stars
3. Star Clusters
4. Nebulae
5. Galaxies

All of the objects listed in this book should be easily seen with a small telescope. By small, I mean something with an aperture of 70mm (about three inches) to 125mm (about five inches.) Specifically, I've tried to list objects that can be seen with a small 70mm 'scope, but it'll be easier if your 'scope is larger.

All of the objects are either easily seen with just your eyes (making them very easy to find) or else they lie very close to a bright star. That way, you'll be able to easily locate them by first finding the star itself. The object will always be within the same finderscope field of view as the star.

Please be aware of one thing: although I've seen the vast majority of these objects from the suburbs of Los Angeles (hello, light pollution) and frequently while they're setting in the west (as opposed to high overhead, where the air is clearest) whether you can see the object will greatly depend upon your location, your equipment and your own eyesight. I've chosen objects that should be visible regardless, but I'm not you, I don't know your personal circumstances and, consequently, I can't guarantee your own personal experiences.

That being said, I'm confident you'll be able to see the vast majority, if not all of them. (A word to readers in the United Kingdom: there are two objects you won't be able to see from your latitude and I apologize in advance for that. These are the Butterfly Cluster and Messier 7. Unfortunately, they're too far south to rise over the horizon and I feel bad about that because I'm English myself. If it's any consolation, they barely skim the southern horizon from Los Angeles while observers in the southern hemisphere are lucky enough to see them pass overhead.)

Before we get started, let's take a few moments to review the information to be found on the individual pages for the objects.

At the top you'll find the object name or identifier. The names of the stars are often Arabic in origin, but you'll notice some may have names like "Gamma Delphini." The first part of the name (gamma, in this example) is a letter of the Greek alphabet. Thousands of years ago, the Greeks assigned letters to the stars based upon their brightness. So the brightest star in the constellation would be assigned the letter Alpha – their equivalent of the letter A. The second brightest would be Beta, then Gamma, Delta and so forth. (There's a table detailing the Greek alphabet in the Appendix on page 165.)

The second part of the name refers to the constellation the star belongs to – in this example, Delphini refers to Delphinus, the Dolphin. (Again, there's a list of constellations in the Appendix on page 168)

The system has been refined and updated since then, but it's still very widely used. After all, not every star in the sky can have a name and this is an easy, convenient alternative.

Some of the other objects (the star clusters, nebulae and galaxies) may have names, such as the Andromeda Galaxy or the Owl Cluster. More often than not, they're named after whatever the object resembles (such as an owl) or sometimes the constellation it resides in (such as Andromeda.)

Charles Messier, circa 1770. Public domain image.

The book also mentions Messier objects. These are objects that were noted by the French comet hunter Charles Messier in the 18th century. Some deep sky objects can appear distinctly comet-like (especially globular star clusters) and Messier wanted to avoid confusing them with any potential discovery.

There are 110 Messier objects in all and this book contains 22 of them. Some are actually quite clearly not comets – for example, the Pleiades – and many astronomers have wondered why Messier went to the trouble of cataloging them. Other objects, such as the famous Double Cluster or the Owl Cluster, are easily visible and yet Messier didn't list them at all.

One other catalog is also well known and well traversed by astronomers – the New General Catalog, or NGC. This list was compiled in the 19th century and contains nearly 8,000 deep sky objects. However, many of them may be beyond the reach of a small 'scope. Incidentally, all of the Messier objects were also assigned an NGC number.

On one side of the object page you'll see three graphics. The first is a general map of the area where the object may be found. The image spans 45° of sky and in the center is a circle. This circle denotes the area of sky seen through a 6x finderscope, as depicted in the second image.

The finderscope image is right-side up; in other words, north is at the top of the picture and south is at the bottom. I mention this because some finders will flip the view; please be aware of this as you're looking for your object. Both these images were created using the *Mobile Observatory* app by Wolfgang Zima. It's available for Android devices and is highly recommended. It's the only astronomy app I use on a daily basis and Wolfgang was gracious enough to let me reproduce the images. (See zima.co for more information.)

The third image on the page attempts to depict the view through your eyepiece. I say *attempts* because I don't know exactly what kind of eyepiece you're using or the field of view you'll get, but this should be a pretty good representation. You'll see the magnification in the top left corner and in the bottom left is an indication of north and east. Your telescope will flip the view; it may just be vertically, or horizontally or both but the view depicted should match many mainstream telescopes.

There are a few other things to bear in mind when looking at these eyepiece views.

With the exception of open clusters, the other deep sky objects are drawn depicting their true size in the sky. In reality, you won't see the object this large in your finder or eyepiece unless you're observing under very dark skies with a large telescope. As stated before, what you see will depend upon your equipment, your location, the sky conditions at the time and your own eyesight. The same is true of some of the background stars depicted; how many you'll see will, again, depend upon those same factors.

Also, it's difficult to accurately depict the multiple stars because, in order to show which stars are brightest, it's necessary to make the dots representing the stars larger. For the sake of clarity, some of the multiple star images may have large magnifications (e.g., 217x) in the corner. In reality, you can probably get a similar view at about half that magnification.

The eyepiece views were created using *Sky Tools 3*, an excellent program for your PC or Mac that allows you to plan an evening under the stars. As you can see, it also accurately simulates the view through your eyepiece. Like *Mobile Observatory*, I consider it invaluable. (See www.skyhound.com for more information.)

My thanks to Greg Crinklaw for his permission to use these images.

You'll notice I don't have any photos of the objects and there's one very good reason for this: more often than not, they don't accurate depict what you'll see through the eyepiece, whereas *Sky Tools* does. Too many new astronomers get discouraged because they see beautiful, colorful images in magazines and online and expect to see something similar. Those photographs are created by dedicated individuals who'll spend hours combining many images and fine tuning the final result until it's the best it can be.

The Whirlpool Galaxy is visible in a small telescope but it requires dark skies and won't look nearly as good as this image! Taken by the author using Slooh. (www.slooh.com)

I hate to break it to you, but your telescope is not called Hubble either. I don't want you to be discouraged; the night sky can be stunning and awe inspiring but sometimes you have to use a little imagination and truly take into account what you're looking at. I've tried to do that in the text that accompanies each object.

On the other side of the page is some information about the object. You'll see its designation (for example, Gamma Andromedae, Messier 45 etc.), the constellation it belongs to and then we have two co-ordinates, R.A. and Declination.

R.A. is short for Right Ascension and this, coupled with Declination, will give you the exact position of the object in the night sky. Think of it this way – if you were to plot the stars on a globe, then R.A. and Declination would be the sky equivalent of longitude and latitude, respectively.

I debated whether to include them as the objects are easily found once you get to know the constellations. However, some of you may be using a star chart, such as the *Pocket Sky Atlas* by Roger W. Sinnott, and anyone with a GoTo can enter the co-ordinates into their hand controller and the telescope will find it for you. (Most of the objects will be included in the GoTo's database anyway.)

Below the co-ordinates you see what type of object it is – for example, multiple stars and open clusters are the most common. I'll talk about these in a moment.

Next you'll see two sets of stars – one for Location and one as a general Rating. In terms of Location, an object with three stars means it's easily visible to the unaided eye. So, for example, a bright star will have three stars for Location.

Objects with two stars for Location are those that can be found by placing a nearby bright star in the finder and the object will appear in the middle. Objects with one star for the Location are those that may be particularly faint or might require you to have the bright star on the edge of the field of view and the object will appear on the opposite edge.

The general rating is more straightforward. I've given a three star rating to the objects I consider to be unmissable and those I always take time to observe. Objects with two stars still provide something worth seeing while those with one star may be appear faint and best observed under clear, dark skies. (As opposed to the light polluted skies of suburbia.)

There are a lot of objects in astronomy that may seem like a "one star" object at first glance, but remember this: there's a sense of accomplishment that comes from being able to track it down and, as you observe it, consider what you're looking at. Again, I've provided some notes to help with this.

(Please also bear in mind that I want you to have realistic expectations of what you'll see. Astronomy is awe-inspiring, challenging and fires the imagination.)

Lastly, I've indicated when the object is best seen. You might be able to see the object at other times of the year but this will sometimes require you to observe during the early hours of the morning. So the seasons listed here indicate when the object is at its best visibility in the evening. (This is also why I included the star charts earlier in the book – so you can see what's up in the evening and immediately start observing.)

Below that you'll see some text about the object. This information is predominantly based upon my own personal notes and experiences, both while observing from Oklahoma and also from Los Angeles. Again, my goal was to provide you with a realistic idea of what you might see from either a suburban or a city location. I've read a lot of books written by astronomers lucky enough to observe under clear, dark skies but I understand that this isn't an option for everyone.

Solar System Objects

Technically, objects within the solar system aren't deep sky objects but I wanted to briefly discuss them as they're well worth taking some time to truly appreciate them. As you might expect, there are many other books on the market that discuss observing the planets and one of the best is *The Planet Observer's Handbook* by Fred W. Price.

The five brightest planets – Mercury, Venus, Mars, Jupiter and Saturn – are visible with just your eyes (although Mercury can be tricky) with Venus, Jupiter and Saturn providing particularly good views through a small telescope. With a little more experience, you'll also be able to spot Uranus and Neptune.

Before you do anything, you'll need to know if any planets are currently visible and this is where the *Mobile Observatory* app or *Sky Tools 3* definitely prove their worth. Alternatively, you can try one of the more basic planetarium programs, such as *Stellarium*, which has a free version available for PC's and Mac's.

Mercury and Venus are both known as *inner planets* because they're closer to the Sun. Consequently, they can never appear opposite the Sun in the sky (unlike the other planets) and can only be seen in the early morning or early evening sky. So, for example, if you're an early riser you might catch them both in the pre-dawn twilight, but if you're looking for them in the evening, you can catch them in the twilight after sunset.

When you're checking their visibility, you'll need to find out when they reach their greatest elongation from the Sun. This is when the planet appears furthest from the Sun in the sky and is well-placed for observation. Here's where it gets slightly confusing. If the planet is at greatest *eastern* elongation it's visible in the evening sky, but in the west. If it's at greatest *western* elongation, it actually means it's visible in the pre-dawn sky, in the east.

There's a table in the Appendix that lists the elongation dates between 2016 and 2025 (see page 166 for details.)

You might need some help finding Mercury as it's often low on the horizon and can be a little faint against the twilight sky. It often appears pink-white and may be twinkling rapidly. Using binoculars will definitely help you. Mercury, like the God it is named after, is fleet-footed and may only be visible for a few weeks at a time.

Venus is a lot easier. In fact, Venus is unmissable. It appears as a brilliant white star, sometimes quite high above the horizon, often for hours after sunset and for months at a time. It's been mistaken for a UFO and provides a beautiful addition to the evening night sky, especially during the holiday season. After the Sun and Moon, it's the third brightest object in the sky.

Through a telescope both planets will show phases like the Moon. Venus is closest and will therefore appear largest with Mercury appearing much smaller. You'll only need about 35x to clearly see the white disc of Venus while about 100x will provide a great view.

Venus, October 6th 2012. Image by the author using Slooh (www.slooh.com)

Mercury's disc is problematic at low power – you'll probably be able to glimpse it at around 35x but you won't be able to clearly see it until you get to around 80x or 90x. To me, it's always appeared a tan or dull gold color.

All the other planets orbit further away from the Sun and can appear opposite the Sun in the sky. When this happens, the planet is said to be at *opposition* and it's the best time to observe the planet as it will rise at sunset and set at sunrise. It's therefore visible all night and will appear at its brightest and largest.

As with Mercury and Venus, there's a table in the Appendix that lists the opposition dates for Mars, Jupiter and Saturn between 2016 and 2025 (see page 167 for details.)

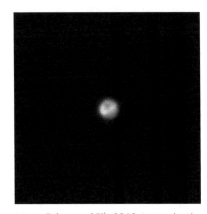

Unfortunately, Mars only reaches opposition once every couple of years and isn't always conveniently placed for observation. Also, the size of its disc will vary, depending upon how close the planet is to us at the time. To the unaided eye it appears bright and coppery and a small telescope will show a tiny disc at low magnification.

At its best (and with higher magnification) you'll be able to see tiny dark markings on its surface and maybe even the glint of an ice cap at the poles. I've seen markings at 54x but you might need to go higher. A magnification of about 100x seems fairly reliable.

Mars, February 25th, 2012. Image by the author using Slooh. (www.slooh.com)

Jupiter and Saturn present, by far, the best views of any planet in the solar system and I promise you will never get bored of viewing them.

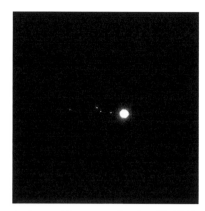

Jupiter and its moons, October 7th, 2012. Image by the author using Slooh. (www.slooh.com)

Many people are wowed by Saturn when they see it for the first time. Personally, I love Jupiter. Why? Two reasons: firstly, because you get to see more and secondly, you can start an observing session with Jupiter and then come back to it just a few short hours later and you'll see that things have changed.

For example, even a pair of steadily held binoculars will reveal the planet's four largest moons – Io, Europa, Ganymede and Callisto. These moons are collectively known as the Galilean satellites after the Italian astronomer Galileo Galilei who first discovered them in January 1610. These moons orbit the planet fairly quickly and can be seen to cast shadows against Jupiter's disc and disappear in eclipses behind the planet.

Beyond these four moons you'll also see dark stripes across the disc of the planet. Typically you'll see two (the northern and southern equatorial belts) at around 40x and if you increase the magnification to around 70x or 80x you may also notice dark patches at the poles.

You probably won't see the Great Red Spot. This is the Earth-sized storm that looks like Jupiter's eye staring out into space in many of the photographs of the planet. Despite many years of looking, I was only able to see it with my 130 SLT with the use of a blue eyepiece filter to help bring out the contrast in the planet's features. Even then, if I hadn't known it was currently visible, I probably would have missed it!

Jupiter, January 4th, 2015. Image by the author using Slooh. (www.slooh.com)

Saturn, of course, is famous for its rings and is usually the planet that beginners want to see first. There's something about a person's first view of the planet that often stays with them and many amateur astronomers can clearly recall when they saw the planet for the first time.

You've probably seen a lot of photos of the planet but when you actually see it for yourself, you'll notice a third dimension that you just can't experience any other way. It looks like a Christmas tree ornament, impossibly hanging there in space. It's easy to understand how the first telescopic observers, some four hundred years ago, were completely confused and bewildered by the rings. Nowadays, thanks to missions like the *Voyager* and *Cassini* space probes, we know the rings to be comprised of millions of chunks of ice, some no bigger than a first, others as large as a house, all orbiting Saturn in unison.

Saturn, December 27th, 2011. Image by the author using Slooh. (www.slooh.com)

Like Jupiter, you'll also be able to see a few moons, most notably Titan, the planet's largest. But unlike the moons of Jupiter, they don't move quickly and the view won't considerably change within a few short hours. However, come back the following night and you'll see some differences.

Beyond that, you may see a gap in the rings (called Cassini's Division, named after the astronomer who discovered it) and maybe some faint markings on the disc, but that's about it. To be honest, for me it's like a childhood friend I've grown apart from. I'll still stop by and visit if it's around, but I don't usually stay for long and I don't find there's much to talk about!

Uranus and Neptune can also be seen but you need to know where to look for them. At low power (say, around 35x) they only appear as tiny, starlike points so even if you're using a GoTo, you'll need to know which are the stars and which is the planet. Having said that, there *is* a way you can tell. Both planets will show a fairly distinct color – a pale turquoise tint for Uranus and a sky blue hue for Neptune.

Uranus might also show a disc; I've possibly seen it at 26x but I've typically needed at least 50x to be sure. Neptune is trickier and I've only seen a very tiny disc at a magnification of 160x or higher.

The same is true for asteroids. You'll need to know exactly what you're looking for and be familiar with the background stars. Unfortunately, unlike the major planets, they're too small to appear as anything but a tiny star-like point in your telescope and can often be disappointing to new astronomers.

One last thing: comets. Occasionally, a reasonably bright one might come close and make itself visible to amateur astronomers with small telescopes. How bright it will be and how it will appear in your 'scope really depends on the comet itself.

No two comets are alike and none are guaranteed to be a spectacular celestial event. I've observed a number and they typically look like misty spheres with a star-like point (the core of the comet) at its center. Some comets will also show a tail while others might have a greenish tint to them.

Be sure to join an astronomical group (be it online or in person) to keep up-to-date with any potential surprise visitors that might be coming around.

Comet C/2013 K5, December 12th, 2012. Image by the author using Slooh. (www.slooh.com)

Multiple Stars

I hate the phrase "star gazing." For me, it always conjures up an image of someone standing still, staring up at the night sky. Or maybe peering through a telescope at a single bright star. Someone else might walk by. "What are you looking at?" they ask. The observer points up at the sky. "That bright star. That one, right there."

Many non-astronomers think this is what "star gazers" do. It's like when I stopped eating white and red meat during the 1990's. People thought I just ate peas and carrots instead. (Incidentally, I've since returned to the dark side.)

The double star Kuma, aka Nu Draconis. Image by the author using Slooh. (www.slooh.com)

Nothing could be further from the truth, especially given that many stars are not single stars at all. In fact, most of them are *multiple* stars. To the unaided eye, they'll appear to be a single star, but when you observe them with binoculars or a telescope, that star is split in two. Some stars may have three or four components.

There are two kinds of multiple stars: those that are true multiple star systems, where the stars orbit one another, and those that are *optical* doubles and only appear close together due to a chance alignment. In reality, they may be light years apart.

Some multiples have components that are equally bright (such as Kuma) while others show stunning colors (such as the famous Albireo.)

The best thing about multiple stars (beside their abundancy and variety) is that they're largely unaffected by moonlight and light pollution. As long as both components are reasonably bright, they'll still shine through the brightened sky caused by an intrusive Moon.

Star Clusters

Messier 41 in the constellation of Canis Major. Image by the author using Slooh. (www.slooh.com)

Star clusters also fall into two categories: open star clusters and globular star clusters. An open star cluster contains tens or hundreds of stars, all of which literally appear clustered together in the same relatively small area of sky. This is not a chance alignment – these stars are genuinely grouped close together in space and are born from the same nebula. Consequently, they're usually quite young and are quite literally, stellar siblings.

Some clusters appear fairly large and only require a low magnification with member stars scattered across the field of view. The famous Pleiades and Praesepe open clusters are excellent examples. Others are small and compact and may require a higher magnification to be properly appreciated. Messier 103 is one example.

The other type of cluster is the globular cluster. These are spherical balls of stars in space and contain thousands of stars within a tightly packed area. They lie thousands of light years away, usually close to the hub of the galaxy, and are very old in comparison to the younger open clusters. It's not unusual for a globular to be over ten billion years old – nearly as old as the universe itself.

Through a small telescope a globular can often appear like the head of a comet without the tail. Alternatively, you might think of it as being a faint and fuzzy star. But increase the magnification and you may be able to resolve some of the individual stars around the outer edges. You may also notice chains of stars or the cluster may appear to be misshapen. Not all globulars are the same! For my money, the best globular in the northern hemisphere is the Keystone Cluster but Messier 22 is a fine example too.

The Keystone Cluster in the constellation of Hercules. Image by the author using Slooh. (www.slooh.com)

Whether you're observing open or globular star clusters, you'll get the best views away from the lights but you can still get some great views from the suburbs, even with light pollution.

Nebulae

There are, of course, several categories of nebulae but the most common are clouds of gas and dust in space. These are the birthplaces of stars and can cover an area light years in diameter. The Orion Nebula is the best example of this in the northern hemisphere. It's easily seen with the unaided eye as a tiny, misty patch in the sword of Orion and can be a stunning sight in a telescope.

Unfortunately, that's the only example of that kind in the book as most nebulae are fainter and don't appear close to a bright star. However, there are a few others that might draw your attention.

The Orion Nebula in the constellation of Orion. Image by the author using Slooh. (www.slooh.com)

One is the Ring Nebula, an outstanding example of a planetary nebula. These nebulae are small and appear disk-like when observed through a telescope, almost like a planet. The Ring is the best and brightest example and literally looks like a tiny smoke ring in space. It's located in the constellation of Lyra and is best seen in the summer.

The other kind of nebula is a supernova remnant. There are only a few of these and only one is easily seen in a small telescope. This is the Crab Nebula and it's the remains of a star that exploded nearly a thousand years ago. The Crab is located in Taurus, the Bull, and is best seen in the autumn and winter.

Galaxies

The Andromeda Galaxy in the constellation of Andromeda. Image by the author using Slooh. (www.slooh.com)

Galaxies come in all different kinds of shapes and sizes but may be disappointing to the beginner. There are a lot of images around showing star-studded spirals in space, but the reality is that you're likely to only see a small, misty patch. Depending upon the object, your equipment and location, you may be able to see some shape and structure and patience definitely pays dividends, but don't get your hopes up if you're affected by light pollution.

Also, the vast majority of galaxies are small, faint and can be difficult to locate. With experience, you'll be able to spot a number of them, even with a small 'scope, but to begin with there's only one that's easily seen. The Andromeda Galaxy appears as a misty patch with the unaided eye and is conveniently located close to a number of bright stars.

(I did also consider including another, the Triangulum Galaxy, but this is harder to locate and can be tricky to see against the background sky.)

The Deep Sky Objects

Pi Andromedae

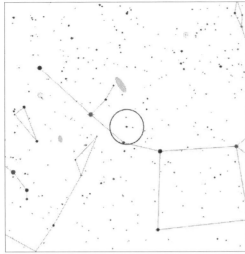
Map courtesy *Mobile Observatory*

Designation(s):	Pi Andromedae
Constellation:	Andromeda
R.A.:	00h 36m 53s
Declination:	+33° 43' 10"
Object Type:	Multiple Star
Location:	★★
Rating:	★★
Best Seen:	Autumn and Winter

Finderscope view courtesy *Mobile Observatory*

Eyepiece view courtesy *Sky Tools 3*

Pi Andromedae is not one of the brightest stars in the constellation but is still quite easily found from suburban or dark sky locations.

Andromeda is always depicted as being a curved line of stars extending east from the square of Pegasus. However, if you look carefully, you'll see a second curved line of fainter stars of which Pi is the first. (The second, Mu Andromedae, is fairly close to the Andromeda Galaxy.)

Binoculars won't split the star but a small telescope at low power should easily do the trick. At 26x I saw a bright white primary with a very faint bluish secondary. From the city I noted that the secondary appeared to have a coppery color, but I suspect that's more likely due to the city air.

There is a third, much fainter and wider companion but I've been unable to see it from either the suburbs or from the city.

The components we see with our telescopes are not members of a true multiple star system but are simply a chance alignment of stars at differing distances from us. However, the primary is actually a very close pair of nearly identical stars that each shine with the light of a thousand Suns. The pair orbit one another once every 144 days and are approximately 600 light years away.

The Andromeda Galaxy

Designation(s):	Messier 31
Constellation:	Andromeda
R.A.:	00h 42m 44s
Declination:	+41° 16' 09"
Object Type:	Spiral Galaxy
Location:	★ ★
Rating:	★ ★
Best Seen:	Autumn and Winter

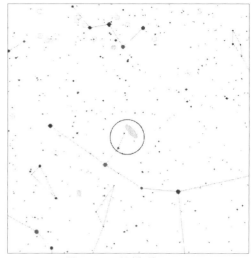
Map courtesy *Mobile Observatory*

Known since antiquity, the Andromeda Galaxy is one of the most fascinating objects you can see – but if you own a small telescope, it can also be one of the most disappointing.

The reason is simple. Although it's visible to the unaided eye under dark skies, its light is spread over a large area and, consequently, it usually doesn't appear that bright (or detailed) through a small telescope.

So what's the fascination? Well, for one thing, you're looking at a whole other galaxy.

Our own Milky Way galaxy has about 200 billion stars and the Andromeda Galaxy is even bigger. And at a distance of 2 ½ million light years, it's the most distant object you can easily see with just your eyes. It's also worth remembering that, as you stare at it through the eyepiece, the light that's now hitting your eyes first started out millions of years ago.

What does this mean? You are, in fact, looking at the galaxy as it once was. You're literally looking into the past.

Under suburban skies at 35x, it appears as a moderately bright, misty oval but in the city it was hard to see at all. In fact, I had to use averted vision to even see its shape. Otherwise, only the core is visible. Look for it, enjoy it, but don't get your hopes up too much!

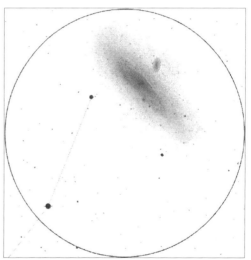
Finderscope view courtesy *Mobile Observatory*

Eyepiece view courtesy *Sky Tools 3*

NGC 752

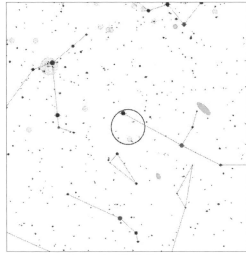
Map courtesy *Mobile Observatory*

Designation(s):	NGC 752
Constellation:	Andromeda
R.A.:	01h 57m 55s
Declination:	+37° 51' 57"
Object Type:	Open Cluster
Location:	★ ★
Rating:	★ ★
Best Seen:	Autumn and Winter

Finderscope view courtesy *Mobile Observatory*

Eyepiece view courtesy *Sky Tools 3*

NGC 752 is one of the larger open clusters and its light is consequently scattered across a relatively large area of the sky.

That being said, it should be visible in binoculars as a faint misty patch and using averted vision may help to resolve some individual stars.

This is one of those clusters that looks good under both suburban and city-based skies. You should be able to locate it with Almach on one edge of the finderscope field of view and the cluster on the opposite edge.

At 35x, it appears as a large, sparse cluster and denser toward the core. The stars are predominantly blue-white but you'll also notice some scattered orange stars throughout.

The double star 56 Andromedae lies on the edge and makes a convenient marker for the cluster. When viewed at this magnification, the entire cluster fits into the field of view with an orange star lying just outside the main group.

On occasion, when observing the cluster as it sets in the west, it takes on the appearance of Boötes with this star taking the place of Arcturus.

Almach

Designation(s):	Gamma Andromedae
Constellation:	Andromeda
R.A.:	02h 03m 54s
Declination:	+42° 19' 47"
Object Type:	Multiple Star
Location:	★ ★ ★
Rating:	★ ★ ★
Best Seen:	Autumn and Winter

Map courtesy *Mobile Observatory*

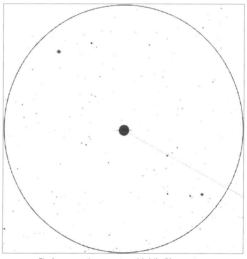

Finderscope view courtesy *Mobile Observatory*

Eyepiece view courtesy *Sky Tools 3*

Almach, the third brightest star in Andromeda, is a true multiple star system located some 350 light years away. Although only two stars can easily be seen with a small telescope, there are, in fact, four stars present with the fainter, blue component being its own triple star system.

Definitely a favorite for the autumn, this double star can be split with low power and is reminiscent of Albireo in Cygnus.

But whereas Albireo is easily resolved at low power, you'll probably need at least 50x to properly enjoy the view. I've barely split it at 26x from the city, but this is one double where greater magnification definitely comes in handy. I've found that a magnification of about 65x or 70x works quite nicely.

The primary is a pale yellow-white gold and about four times brighter than the pale blue secondary. On a few occasions, I've noted that the secondary will flash purple and violet, but these colors may only be apparent at higher magnification (around 90x or 100x.)

One last thing – I've noticed that Almach is one of those pairs where the colors were more apparent at higher magnifications. What do you see?

Almach was first seen as a double in 1778 by the German physicist Johann Tobias Mayer.

Achird

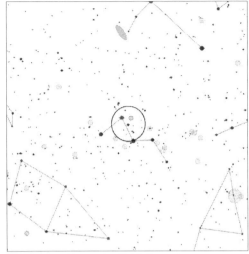
Map courtesy *Mobile Observatory*

Designation(s):	Eta Cassiopeiae
Constellation:	Cassiopeia
R.A.:	00h 49m 06s
Declination:	+57° 48' 55"
Object Type:	Multiple Star
Location:	★ ★ ★
Rating:	★ ★ ★
Best Seen:	Autumn and Winter

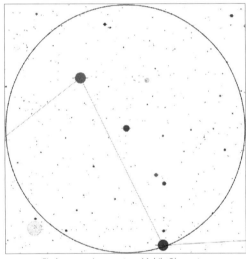
Finderscope view courtesy *Mobile Observatory*

Eyepiece view courtesy *Sky Tools 3*

Achird is a relatively close multiple star; many sources will mention a gold primary with a reddish-purple companion, but I've noted other colors and a third, bluish star nearby.

Although it doesn't require much power to reveal the two main components, I've been unable to split it at a magnification of 26x but upping it a little to 35x or 40x should do the trick.

The first, and brightest, component appears a brilliant white to me (rather than the gold noted by others) but you'll need to focus carefully to reveal the much fainter companion.

At this magnification, it's difficult for me to notice any particular color but increasing the magnification again, to about 50x, helps to reveal the companion's coppery complexion.

While you have it at about 50x, try looking for the third component, which should appear as a tiny blue star on the opposite side of the white primary.

If not, increase the magnification once again. I've been able to spot it at 65x but 87x seemed to provide a particularly nice view with all three stars visible.

The Owl Cluster

Designation(s): NGC 457

Constellation: Cassiopeia

R.A.: 01h 19m 33s

Declination: +58° 17′ 27″

Object Type: Open Cluster

Location: ★

Rating: ★ ★ ★

Best Seen: Autumn and Winter

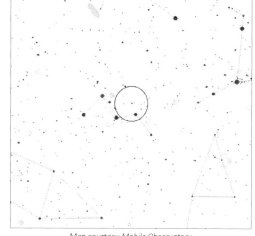
Map courtesy *Mobile Observatory*

A definite favorite, this cluster has a very distinctive shape that has given rise to a number of imaginative names over the years. Many see the stars forming the shape of an owl, while others see a kite. More recently, some have come to call it the E.T. Cluster.

Although the cluster can be seen with binoculars it doesn't truly shine until you turn your telescope toward it. You won't need a lot of magnification either – it can be seen with 26x and 35x provides a nice view, but somewhere between 50x and 90x is probably best. Once you get to about 100x, you'll have difficulty fitting the entire cluster into the field of view.

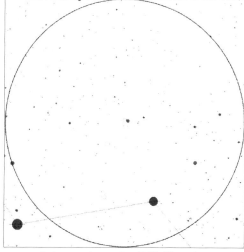
Finderscope view courtesy *Mobile Observatory*

The first time you see it, you're sure to be delighted as your eyes take in the sight and your imagination takes over. With the double star Phi Cassiopeiae marking the eyes, it's easy to see the cluster as either an owl with powerful wings or an alien with long, outstretched arms staring back at you.

The two stars of Phi Cassiopeiae are both white, with one being about 1½ times brighter than the other. It's a sparsely scattered cluster with the densest portion being around the chest area of the alien.

If the thought of an alien spooks you, look again. No owl? How about a goose or a swan in flight with Phi marking the tail?

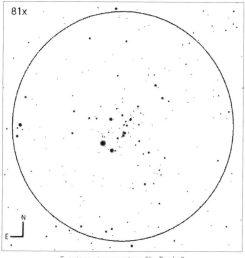
Eyepiece view courtesy *Sky Tools 3*

Messier 103

Map courtesy *Mobile Observatory*

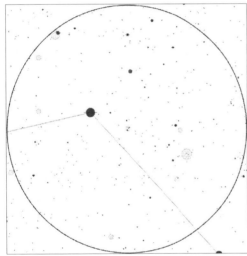
Finderscope view courtesy *Mobile Observatory*

Eyepiece view courtesy *Sky Tools 3*

Designation(s):	Messier 103
Constellation:	Cassiopeia
R.A.:	01h 33m 22s
Declination:	+60° 39' 29"
Object Type:	Open Cluster
Location:	★
Rating:	★
Best Seen:	Autumn and Winter

Messier 103 is a small open cluster that's easily found but may be a little disappointing under city and suburban skies. It's barely visible as a tiny, compact elongated patch in binoculars but you may still be able to see the three stars that make up the slightly curved line across the center.

The telescopic view doesn't improve much from the city. I was only able to see about four or five stars at 26x – the central three for sure – with the center star appearing reddish.

The cluster fares better under suburban skies but it's best to observe it with low expectations. At 35x I could easily see the three bright stars across the center, like a mini Orion's belt. The middle star appeared to be golden with a blue companion close by while the other stars appeared to be white. I also noticed a triangle of stars to the north-east.

Upping the magnification to 91x definitely helped. The cluster still fitted within the field of view and I was able to see a lot more stars with averted vision.

There's one simple reason why the cluster appears so small: at a distance of about 9,000 light years, Messier 103 is one of the most distant open clusters known. There are thought to be over 150 member stars with the cluster being about 25 million years old.

NGC 663

Designation(s):	NGC 663
Constellation:	Cassiopeia
R.A.:	01h 46m 16s
Declination:	+61° 13' 06"
Object Type:	Open Cluster
Location:	★ ★
Rating:	★ ★ ★
Best Seen:	Autumn and Winter

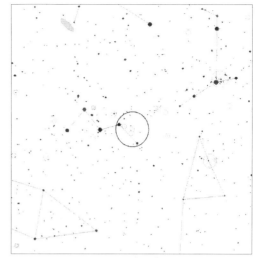

Map courtesy *Mobile Observatory*

NGC 663 is a little gem of a cluster. Like the Owl Cluster, it's easily found and is quite large and attractive, which leads you to wonder why Messier didn't include it in his famous list.

This cluster can be easily seen with binoculars as a large, faint, hazy circular patch about twice the size of Messier 103. It looks like a sparsely scattered globular cluster and you may be able to see individual stars with averted vision.

However, it's best observed with a telescope at low to medium power, somewhere between 50x and 100x.

At 35x it appeared to be a sparsely scattered collection of stars that formed a diamond shape. The vast majority of stars were blue-white, but there were two pairings of stars – one just off center and the other near the edge – that appeared to be orange.

The two pairs of stars are actually a little spooky and make me think of two pairs of eyes. With the cluster high in the sky at Halloween, maybe they're ghosts. Or if NGC 457 is the E.T. Cluster, maybe E.T. is just the largest and friendliest of the aliens and NGC 663 are a smaller pair following close behind.

The cluster lies about 7,000 light years away and contains some 400 stars.

Finderscope view courtesy *Mobile Observatory*

Eyepiece view courtesy *Sky Tools 3*

Mesarthim

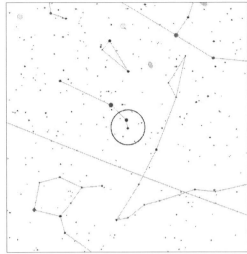
Map courtesy *Mobile Observatory*

Designation(s):	Gamma Arietis
Constellation:	Aries
R.A.:	01h 53m 32s
Declination:	+19° 17' 38"
Object Type:	Multiple Star
Location:	★ ★ ★
Rating:	★ ★ ★
Best Seen:	Autumn and Winter

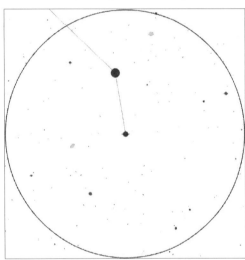
Finderscope view courtesy *Mobile Observatory*

Mesarthim is a popular and relatively easy double star, known to many amateur astronomers across the world.

It's easily found and a good target if you're new to the hobby, especially as it can be used as a good benchmark for your equipment and the seeing conditions.

For example, you should be able to split the star at a fairly low magnification. From the city, I can barely split it at 26x with my 130mm 'scope but at 35x it's more apparent. I usually had no trouble splitting the star at that magnification from the suburbs with my 4 ½ inch XT.

How cleanly the star is split can depend upon the atmosphere; if the air isn't steady, it might appear to be barely split but otherwise it should be pretty clean.

You'll want to increase the magnification to get the best view; somewhere around 100x is probably best but even at lower magnification you should be able to see a third, unrelated fainter star nearby. What color is that star? To me it appears blue.

There's no mistaking the color of the Mesarthim stars themselves as both components are pure white and of equal brightness. Discovered by Robert Hooke in 1664, the two stars orbit one another every 5,000 years and lie about 165 light years away from us.

Eyepiece view courtesy *Sky Tools 3*

Lambda Arietis

Designation(s):	Lambda Arietis
Constellation:	Aries
R.A.:	01h 57m 56s
Declination:	+23° 35' 46"
Object Type:	Multiple Star
Location:	★ ★ ★
Rating:	★ ★ ★
Best Seen:	Autumn and Winter

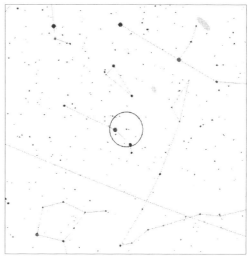

Map courtesy *Mobile Observatory*

Lambda Arietis is slightly off the beaten track but should still be quite easily found. It lies between two brighter stars – Hamal to the east and Sheratan to the west. Lambda appears slightly closer to Hamal in the map and finderscope views depicted here.

Some observers have reported this as being split with binoculars, but I've had no luck with my 10x50's.

However, a low magnification of 26x should easily reveal a brilliant white star that's about twice as bright as the bluish secondary. (On one occasion I noted hints of violet in the companion but the color seemed to disappear with a higher power eyepiece.)

A magnification of about 65x will provide a nice view. Look out for a third, slightly fainter companion nearby that forms an elongated triangle with the main pair.

Lambda is thought to be a true double star system as both stars appear to be moving together through space. In reality, both stars are actually yellow but the primary is hotter and its color is closer to yellow-white.

The reason the secondary appears blue is because your eyes are sensitive to contrasts in light and you see an exaggerated contrast effect as a result.

The stars lie at a distance of about 129 light years and take more than 33,000 years to orbit one another.

Finderscope view courtesy *Mobile Observatory*

Eyepiece view courtesy *Sky Tools 3*

The Double Cluster

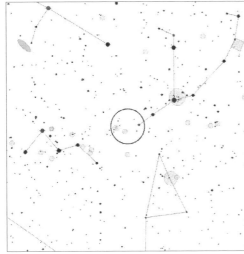
Map courtesy *Mobile Observatory*

Designation(s):	NGC 869 & NGC 884
Constellation:	Perseus
R.A.:	02h 22m 29s
Declination:	+57° 09' 00"
Object Type:	Open Cluster
Location:	★ ★
Rating:	★ ★ ★
Best Seen:	Autumn and Winter

Finderscope view courtesy *Mobile Observatory*

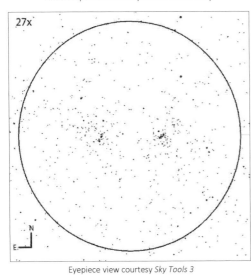
Eyepiece view courtesy *Sky Tools 3*

Known since antiquity, the Double Cluster is one of the highlights of the night sky and yet, somehow, like NGC 663, Messier never included it in his catalog. They are, quite simply, unmissable in every sense.

Depending on the quality of your sky, you may be able to glimpse this stellar pair with just your eyes. Look about midway between the top of Perseus, the Hero and Cassiopeia, the Queen and you might see a faint, elongated patch of hazy light.

If you don't have any luck, try scanning the area with binoculars or your finderscope as they should be easily seen in both. To me, in my small 8x30 binoculars, the clusters seemed to take on the form of a butterfly or bowtie.

These clusters appear fairly large in the sky and are consequently best viewed at low power but I've been able to fit them both into the same field of view up to about 70x or so.

NGC 884, the eastern cluster (to the left in the eyepiece depiction) is larger and more sparsely scattered than its neighbor. NGC 869 has a dense core with two bright blue-white stars, just off-center. Take your time with these clusters and allow your eyes to soak in the view. Most of the stars will be young and blue-white in color but you'll see a few older orange giants too.

Messier 34

Designation(s):	Messier 34
Constellation:	Perseus
R.A.:	02h 42m 07s
Declination:	+42° 44' 46"
Object Type:	Open Cluster
Location:	★ ★
Rating:	★ ★
Best Seen:	Autumn and Winter

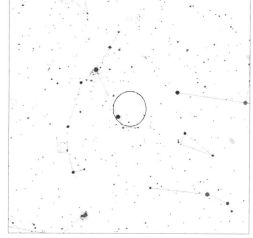
Map courtesy *Mobile Observatory*

Messier 34 is like a cousin to the Double Cluster and is often overlooked in favor of its flashier neighbors.

It can be found by pointing your finderscope or binoculars midway between Almach in Andromeda and Algol in Perseus. By placing Algol on the edge of the finderscope view, you should be able to see M34 on the opposite edge.

This is a fairly large, sparse cluster that reminds me of the Praesepe (M44) and is best viewed at low power. I found a magnification of about 50x to be best.

At 35x I noted that the cluster appeared to be comprised almost entirely of blue-white stars but with a few that show hints of amber. In particular, I've noticed a golden star on its western edge.

There are also a number of apparent double stars scattered throughout with a pair of equal brightness near a triangle at the center.

Do you notice any shapes? I saw an elongated X at 26x and I've also noted a pattern that's reminiscent of the Owl Cluster at the heart of the cluster.

The cluster lies about 1,500 light years away and is thought to contain close to 500 stars.

Finderscope view courtesy *Mobile Observatory*

Eyepiece view courtesy *Sky Tools 3*

Polaris

Map courtesy *Mobile Observatory*

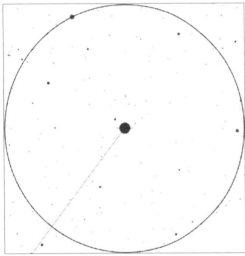
Finderscope view courtesy *Mobile Observatory*

Eyepiece view courtesy *Sky Tools 3*

Designation(s):	Alpha Ursa Minoris
Constellation:	Ursa Minor
R.A.:	02h 31m 49s
Declination:	+89° 15' 51"
Object Type:	Multiple Star
Location:	★ ★ ★
Rating:	★
Best Seen:	All Year Round

It's a common misconception among non-astronomers that Polaris, the pole star, is actually the brightest star in the sky. This has always baffled me as I can't imagine how this belief may have come about.

It's actually just a star of average brightness but it's very easily found and, since it's almost exactly at the north celestial pole, it never moves in the sky. (At least, not to the unaided eye.) It therefore has the convenience of being visible on every clear night of the year.

To be honest, it's not a very exciting multiple star but it's definitely one that should be on every newbie's bucket list. I haven't been able to split it at 26x and you may have to almost triple that power to about 75x to see the companion.

You should see a white primary with a very faint bluish secondary. It's an easy split at 130x but the secondary was difficult to see from the city and I needed to use averted vision to make it more apparent.

Once it's been split and you know where to look for the companion, gradually lower the magnification and see how low you can go before it disappears. Under suburban skies I've definitely seen it at 54x and have suspected the secondary at 35x.

The Pleiades

Designation(s):	Messier 45
Constellation:	Taurus
R.A.:	03h 47m 58s
Declination:	+24° 09' 50"
Object Type:	Open Cluster
Location:	★ ★ ★
Rating:	★ ★ ★
Best Seen:	Autumn and Winter

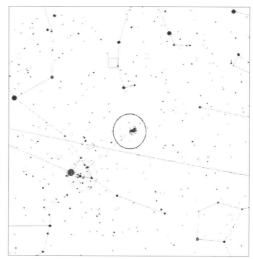
Map courtesy *Mobile Observatory*

Astronomically speaking, winter is a great time for observers in the northern hemisphere. Not only can we admire the beautiful Orion Nebula, but we're also treated to the Pleiades, a stunning open star cluster in the constellation of Taurus the Bull.

It's a famous cluster that's been known by numerous names across the world for many, many years. In Japan, it's known as the Subaru (hence, the car manufacturer) while elsewhere it's known as the Seven Sisters. And this is something of a mystery because although the cluster is easily seen with just your eyes, most observers can only count six stars at most. So what happened to the seventh?

It's a large cluster (it appears twice the size of the full Moon in the sky) and, consequently, some say it's best observed with binoculars where it can be appreciated against the background sky.

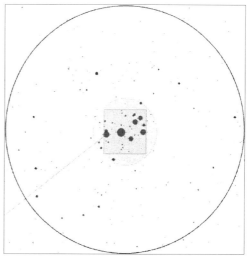
Finderscope view courtesy *Mobile Observatory*

However, if you turn a small telescope toward it with a low magnification eyepiece, you won't be disappointed. The entire cluster should fit into the field of view at about 35x with hundreds of blue-white stars scattered across the scene. Increasing the magnification might actually ruin the effect.

Lastly, if you get the opportunity, observe this cluster well away from any light pollution. You won't regret it.

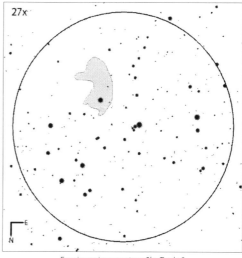
Eyepiece view courtesy *Sky Tools 3*

The Crab Nebula

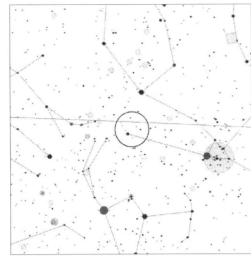
Map courtesy *Mobile Observatory*

Designation(s):	Messier 1
Constellation:	Taurus
R.A.:	05h 34m 32s
Declination:	+22° 00' 52"
Object Type:	Supernova Remnant
Location:	★
Rating:	★
Best Seen:	Autumn and Winter

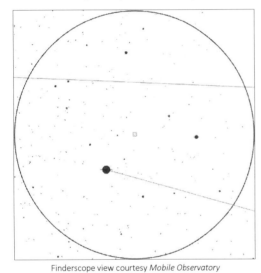
Finderscope view courtesy *Mobile Observatory*

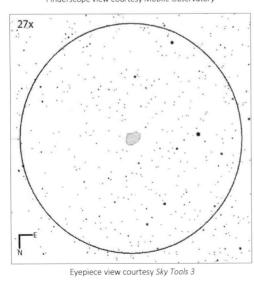
Eyepiece view courtesy *Sky Tools 3*

The Crab Nebula is thought to be the remains of a star that exploded almost a thousand years ago. On July 4th, 1054 Chinese astronomers recorded seeing a "guest star" in the sky. This supernova was so bright it was visible during daylight hours for about three weeks.

Nearly seven hundred years later, in 1731, the nebula was discovered by the astronomer John Bevis in the same area of the sky, but it wasn't linked to the supernova until the early 20th century.

To be honest, it can be disappointing when viewed in a small telescope, but it's worth remembering what you're looking at. There aren't too many supernovae remnants around and the Crab is the only example that can be readily observed by amateurs with inexpensive equipment.

Some can spot the nebula with binoculars but I've not had such luck and I suspect you'll need clear dark skies to be successful. However, it should be visible under suburban skies with a small telescope.

It's barely seen at about 35x but it's not an easy thing to see. Using averted vision, it appears as a very faint oval patch with little or no brightening near the center. I once described it as "a coal stain on a black carpet." Increasing the magnification may help, but if you live in the city, you'll probably be out of luck as I've never seen it from Los Angeles.

Mintaka

Designation(s):	Delta Orionis
Constellation:	Orion
R.A.:	05h 32m 00s
Declination:	-00° 17′ 57″
Object Type:	Multiple Star
Location:	★ ★ ★
Rating:	★ ★
Best Seen:	Winter

Map courtesy *Mobile Observatory*

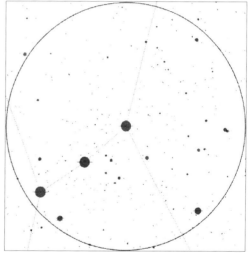

Finderscope view courtesy *Mobile Observatory*

Eyepiece view courtesy *Sky Tools 3*

Mintaka is a bright star, easily identified as being the most westerly star in Orion's belt. The belt has been a famous astronomical fixture for thousands of years with the three stars featuring prominently in cultures around the world.

But before you point your 'scope in its direction, scan the area with binoculars or a small telescope at low power. What can you see? The three belt stars are actually members of a very large open star cluster, known as the Orion OB1 Association and the area is studded with stars.

Mintaka itself is a fairly wide double, but unfortunately not so wide that you can split it with binoculars. (Or at least I've been unable to.)

Having said that, you won't need a lot of power to split it as about 26x should do the trick. At this magnification the two components are easily seen, with the white primary appearing about three or four times brighter than the sky blue secondary.

It looks better at higher magnification with about 50x providing a good view. Something I've noticed is that the higher the power, the stronger the colors appeared to be. Unfortunately, the pair may be less visually appealing as the gap between them appears to increase.

Meissa

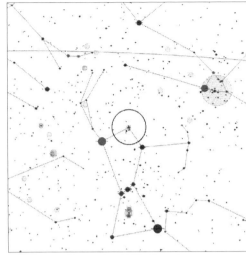
Map courtesy *Mobile Observatory*

Designation(s):	Lambda Orionis
Constellation:	Orion
R.A.:	05h 35m 08s
Declination:	+09° 56' 03"
Object Type:	Multiple Star
Location:	★ ★ ★
Rating:	★ ★ ★
Best Seen:	Winter

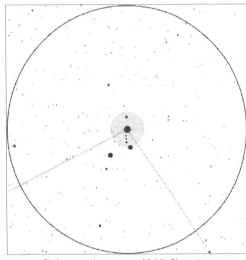
Finderscope view courtesy *Mobile Observatory*

Eyepiece view courtesy *Sky Tools 3*

Meissa is the brightest member of a small cluster of stars that mark Orion's head. Through binoculars or a finderscope you should see a triangular asterism with a line of three faint stars close to Meissa itself.

At 26x, you'll see Meissa as a white star with a much fainter secondary nearby. If you look carefully with averted vision, you might be able to catch a third star between the two.

Up the power to about 35x and the third star should be easily seen, helping to form another line of three stars, equally spaced, with Meissa being at the end of the line and much brighter than the other two.

But that's not all. Increase the magnification again to about 70x and you might just be able to split Meissa itself. Stare for a few seconds and wait for the air to steady and a violet-blue companion will appear to be touching the star.

Meissa is one of those stars where a higher magnification will definitely benefit you. You'll need about 100x to cleanly split the pair (you might even see another, faint blue star nearby at this magnification) but the higher you go, the better the view.

Use your barlow!

The Orion Nebula

Designation(s):	Messier 42
Constellation:	Orion
R.A.:	05h 35m 17s
Declination:	-05° 23' 28"
Object Type:	Nebula
Location:	★ ★ ★
Rating:	★ ★ ★
Best Seen:	Winter

Map courtesy *Mobile Observatory*

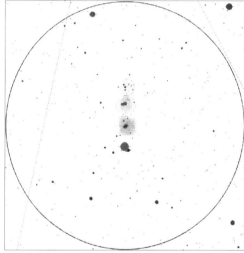

Finderscope view courtesy *Mobile Observatory*

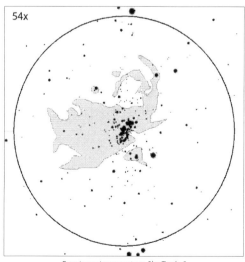

Eyepiece view courtesy *Sky Tools 3*

By far the brightest and best nebula visible from the northern hemisphere, the Orion Nebula can be described with a single word: stunning.

Easily seen with just the unaided eye, it'll look good even through binoculars or a halfway decent finderscope. The cloud appears small and misty, but you should be able to pick out three or four bright stars in the heart of the nebula. These stars, known as the Trapezium, are young stars recently born from the nebula and still swaddled in the surrounding stellar nursery.

Although best observed with low or medium magnification, the nebula stands up to quite a bit of power with greater magnifications revealing more details to the observer. Unlike some celestial sights, you'll notice some color here too – observers often report a greenish tint with the nebula appearing smoke-like. I've noted that it looks as though it's illuminated by moonlight.

While you're here, look out for a small, circular misty patch nearby with a lone star within it. This is Messier 43 (M43), a fragment that appears separated from the main nebula itself. You'll also see a darker, triangular shape on that edge, close to the Trapezium and M43. This is known as the Fish's Mouth and appears as a result of darker matter that obscures the light of the nebula beyond.

The Orion Nebula is about 1,300 light years away and some 12 light years in diameter.

Sigma Orionis

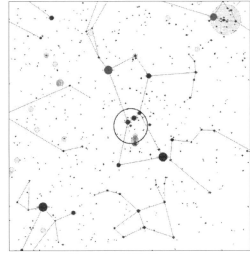
Map courtesy *Mobile Observatory*

Designation(s):	Sigma Orionis
Constellation:	Orion
R.A.:	05h 39m 34s
Declination:	-02° 35' 36"
Object Type:	Multiple Star
Location:	★ ★ ★
Rating:	★ ★ ★
Best Seen:	Winter

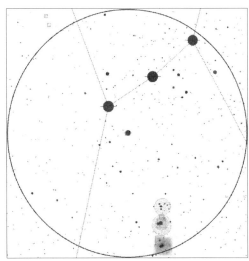
Finderscope view courtesy *Mobile Observatory*

Eyepiece view courtesy *Sky Tools 3*

Sigma is quite a fascinating multiple star that's easily found just below Orion's belt. What's particularly interesting is that it appears close to a mini belt at 26x.

I've also noted that it appears in the middle of an asterism that resembles the constellation Sagitta, with two stars to the west that form the feathers of the arrow. The southernmost star is itself a double with both components being of equal brightness.

Sigma itself has three components with the primary being a brilliant white star. There is a secondary, wide companion to the north-east that appears to be white and about 1½ times fainter than the primary. Look out for a slightly fainter third star, much closer than the second, which is also white and just east of the primary.

The system lies about 1,150 light years away and actually comprises of five stars. The primary you see in a telescope is a very close pair of young white dwarves that orbit one another once every 170 years.

The next star in the system is just to the west of the primary but will be too close and faint to be seen with small 'scopes from suburban or city skies.

The remaining two stars are also dwarves about seven times the mass of the Sun with one being a little-understood, helium-rich magnetic star.

Gamma Leporis

Designation(s):	Gamma Leporis
Constellation:	Lepus
R.A.:	05h 44m 28s
Declination:	-22° 26' 54"
Object Type:	Multiple Star
Location:	★ ★ ★
Rating:	★ ★
Best Seen:	Winter

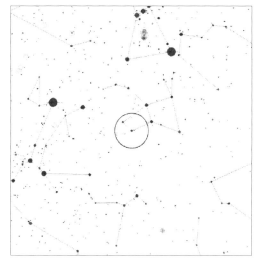
Map courtesy *Mobile Observatory*

Although not split with regular binoculars, this is a fairly wide double that should reveal both the two brightest components at a low magnification.

At 26x, the primary appears to be a creamy white and about twice as bright as the coppery-orange secondary. (On one occasion, I noted the companion appeared to have a blue-purple tint.)

Things get a little more interesting when you up the power. If you increase the magnification to about 90x, you might be able to see a third, fainter star, flickering in and out with averted vision. It appears about three times further away than secondary.

Go a little further – about 100x or even a little higher – and you might catch a fourth star. It's very faint and appears on the opposite side of the primary from the second star. If you look at the distance between the primary and secondary stars, the fourth should appear a little further away.

At just under thirty light years, this system is relatively close to the Earth and is interesting for one other major reason: the primary star is Sun-like (although only about half the age) and may have Earth-sized planets orbiting it. Something to consider as you stare across space at this fascinating star.

Finderscope view courtesy *Mobile Observatory*

Eyepiece view courtesy *Sky Tools 3*

Messier 37

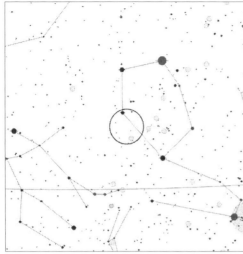
Map courtesy *Mobile Observatory*

Designation(s):	Messier 37
Constellation:	Auriga
R.A.:	05h 52m 18s
Declination:	+32° 33' 02"
Object Type:	Open Cluster
Location:	★
Rating:	★★★
Best Seen:	Winter

Finderscope view courtesy *Mobile Observatory*

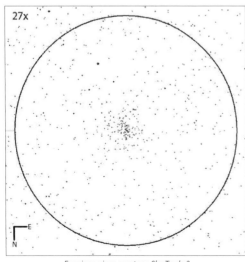
Eyepiece view courtesy *Sky Tools 3*

Messier 37 is the brightest of three Messier open star clusters in Auriga and, lying close to Theta Aurigae, is also the easiest found.

With Theta on the edge of your finderscope field of view, look for the cluster close to the opposite edge. It should appear as a very faint, misty patch under suburban skies.

The other two clusters, Messier 36 and Messier 38, lie to the west, in the south-eastern portion of Auriga's misshapen hexagon and are also worth seeking out.

It's a rich cluster (the richest of the three) and should be easily observed from suburban skies but may be more problematic from the city.

It presents a very nice view at 35x and I noted that there seemed to be a thousand or more stars in the cluster. Using averted vision, it seemed as though a thousand more would spring into view. I've also noted a number of triangular patterns within the cluster. You should be able to increase the magnification to about 50x and still fit the cluster within the field of view.

It's definitely a cluster that begs to be stared at. Just allow your eyes to take in the view, enjoy it and consider that it lies about 4,500 light years away and its 500 stars are scattered across some 25 light years of space.

Messier 35

Designation(s):	Messier 35
Constellation:	Gemini
R.A.:	06h 08m 56s
Declination:	+24° 21' 28"
Object Type:	Open Cluster
Location:	★ ★
Rating:	★ ★ ★
Best Seen:	Winter

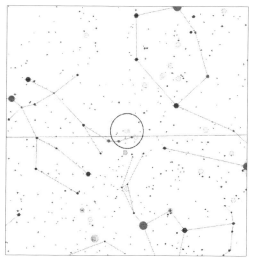

Map courtesy *Mobile Observatory*

Messier 35 is a large, bright open star cluster that should be easily found with your finderscope – or even with just your unaided eyes.

Through binoculars, it appears to be a faint, grey misty patch of uniform brightness but with an hourglass shape. If you look with averted vision, the patch may appear to be resolved into hundreds of tiny points of light – the cluster's individual stars.

Given its size, this is a cluster that's best observed at lower power and about 30x should do the trick. At this magnification, the cluster nicely fits into the field of view and appears to be surrounded by a background scattering of stars.

It's fairly sparse but with plenty of faint stars. Look for a pale gold star on the edge of the cluster with a blue star close to it, giving it the impression of a double star. The golden star is about 1½ times brighter than its blue companion.

Of the others, the remaining stars appear to be either blue or blue-white.

The cluster was discovered in 1745 and lies at a distance of 2,800 light years from Earth.

Finderscope view courtesy *Mobile Observatory*

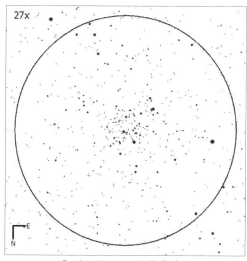

Eyepiece view courtesy *Sky Tools 3*

Castor

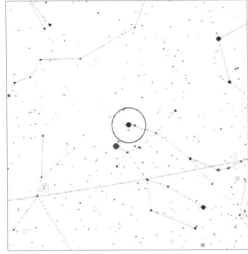
Map courtesy *Mobile Observatory*

Designation(s):	Alpha Geminorum
Constellation:	Gemini
R.A.:	07h 34m 36s
Declination:	+31° 53' 18"
Object Type:	Multiple Star
Location:	★ ★ ★
Rating:	★ ★ ★
Best Seen:	Winter

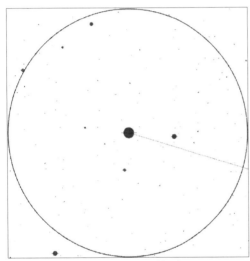
Finderscope view courtesy *Mobile Observatory*

Eyepiece view courtesy *Sky Tools 3*

Castor, despite being designated as Alpha Geminorum, is actually the second brightest star in the constellation of Gemini, the Twins. To the unaided eye, it appears to be a solitary, bright white star but in reality it's a quadruple star system located some 51 light years away.

Unfortunately, only two of the four can be seen with amateur equipment and a magnification of about 50x is needed to split it.

At 26x you'll probably notice two very faint companions of almost equal magnitude within the same field of view. Although this is interesting in itself, this isn't what you're here to see.

Try upping the magnification to about 50x and take another look. Does Castor appear to be split? On some nights, I've barely split it at 54x. On other occasions, I can't say it's been resolved. A lot will depend upon your equipment and the sky conditions.

Double the magnification to about 100x and the view improves immensely. Castor should now be closely split into a pair of brilliant white stars, of almost equal magnitude. Through the 130mm 'scope, there also appeared three fainter stars visible nearby, equally spaced out like Orion's belt but with the middle star a little out of line from the others.

Don't miss this winter wonder!

Beta Monocerotis

Designation(s):	Beta Monocerotis
Constellation:	Monoceros
R.A.:	06h 28m 49s
Declination:	-07° 01' 59"
Object Type:	Multiple Star
Location:	★ ★
Rating:	★ ★
Best Seen:	Winter

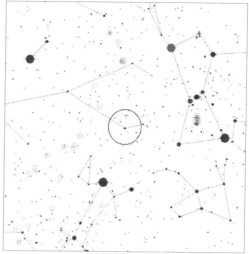
Map courtesy *Mobile Observatory*

Beta Monoceros is a true triple star system some 700 hundred light years away. Two of the stars should be a relatively easy target for a small 'scope and a low magnification eyepiece.

It's barely split at 26x but upping the magnification to 35x or more should make it cleaner and more obvious.

You'll see a pair of white stars of equal brightness with a third, fainter star nearby. (My estimates of the third star's brightness seem to vary somewhat.)

Increasing the magnification might reveal some color (pale yellow and pale blue,) but I only noticed this from the city so there's a real chance it had more to do with the quality of the air than the stars themselves.

Finderscope view courtesy *Mobile Observatory*

You may need to push your 'scope to the limit to see all three stars. If you have a barlow, increase your magnification to as far as your 'scope can usefully go and try your luck.

At 217x, with the 150mm, I noted that one of the two brightest stars appeared to be elongated and, therefore, potentially multiple, but this was at the limit of the telescope's maximum magnification and I was unable to actually split them.

What do you see?

Eyepiece view courtesy *Sky Tools 3*

Sirius

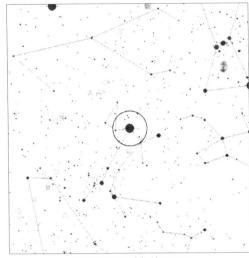
Map courtesy *Mobile Observatory*

Designation(s):	Alpha Canis Majoris
Constellation:	Canis Major
R.A.:	06h 45m 09s
Declination:	-16° 42' 58"
Object Type:	Multiple Star
Location:	★ ★ ★
Rating:	★
Best Seen:	Winter

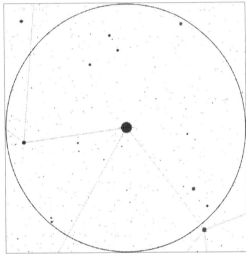
Finderscope view courtesy *Mobile Observatory*

Eyepiece view courtesy *Sky Tools 3*

Sirius, as many people know, is the brightest star in the entire night sky and has been known throughout history by many different civilizations and cultures.

It's a brilliant blue-white star that will often sparkle and flash a myriad of colors when close to the horizon. In fact, its name is actually derived from the Greek for "glowing" and it's not unusual for unsuspecting folk to report it as a UFO.

Easily found by drawing a line through Orion's belt down and toward the east, it's a very easy target and you may have already turned your telescope toward it. What did you see? Personally, I was very surprised to find that it can be dazzling when observed, even at low power.

At 26x I've noticed that it forms a triangle with two much fainter stars and there's a fourth, coppery star between those two that's fainter than all of them.

Sirius is a little over eight and a half light years away and is one of the closest stars to the Earth. Besides the primary star itself, there's a smaller white dwarf companion that can be glimpsed with large telescopes.

This companion was once a red giant that slowly died and shrunk about 120 million years ago. As the primary, Sirius A, is often known as the "dog star" the white dwarf companion, Sirius B, is sometimes nicknamed "the pup."

Messier 41

Designation(s):	Messier 41
Constellation:	Canis Major
R.A.:	06h 46m 00s
Declination:	-20° 45' 15"
Object Type:	Open Cluster
Location:	★★
Rating:	★★★
Best Seen:	Winter

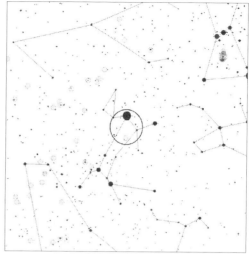
Map courtesy *Mobile Observatory*

Messier 41 is one of my favorite clusters, for a number of reasons. Firstly, it's very easily located, thanks to the proximity of Sirius. All you need to do is point your finder toward that bright star and with Sirius near the top, the cluster will appear near the bottom.

Through binoculars under suburban skies, it appears quite large and has a definite shape to it. It looks as though there's a circular hole on the western edge of the cluster that gives it the appearance of a lobster with its claws outstretched.

That impression is reinforced when the cluster is observed at 35x. I feel it's one of the best (and under-rated) clusters in the sky, one that allows your imagination to run wild.

Finderscope view courtesy *Mobile Observatory*

It's a nice, large cluster with sparsely scattered stars that reminded me of NGC 457 in Cassiopeia, only with a lot more stars. The vast majority are white or blue-white and many are of about the same brightness. However, there's a pair of stars, just a little brighter than the others and another with an orange hue. This reminds me of Phi Cassiopeiae.

The other stars in the cluster are grouped - to me - as though they're forming the shape of a man with his hands up, or the afore-mentioned lobster... or wings like an angel... or a dove... what do you see?

Eyepiece view courtesy *Sky Tools 3*

Procyon

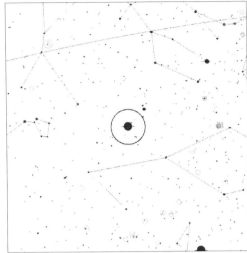
Map courtesy *Mobile Observatory*

Designation(s):	Alpha Canis Minoris
Constellation:	Canis Minor
R.A.:	07h 39m 18s
Declination:	+05° 13′ 29″
Object Type:	Multiple Star
Location:	★★★
Rating:	★
Best Seen:	Winter

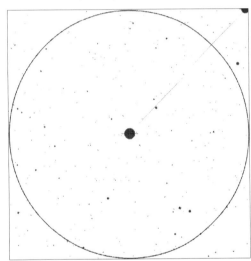
Finderscope view courtesy *Mobile Observatory*

Eyepiece view courtesy *Sky Tools 3*

Procyon is a bright star, easily located to the east of Orion. Use the two stars of his shoulders - Betelgeuse and Bellatrix - to help identify it. (See page 31.)

Through the finderscope (or binoculars) it appears as a brilliant white star with a wide, faint blue-white companion nearby. This companion appears about ten times fainter than Procyon itself.

The pair will still appear within the same field of view at 35x but you'll also notice a scattering of many fainter stars in the background. In particular, look out for another faint star near Procyon and two others near the companion.

Procyon is the brightest of only two bright stars in the constellation of Canis Minor, the Little Dog. As with Sirius, the brightest star in Canis Major, the Great Dog, it follows Orion across the sky, just as the two dogs follow their master as he embarks on his winter hunting trips.

(They all seem totally oblivious to their prey, Lepus the Hare, which can be found below Orion. Maybe they've been distracted by Taurus the Bull, charging from the west.)

Again, like Sirius, Procyon is another bright, white star with a white dwarf companion that's just under eleven and a half light years away. The pair orbit one another every forty years.

Messier 93

Designation(s):	Messier 93
Constellation:	Puppis
R.A.:	07h 44m 29s
Declination:	-23° 51' 11"
Object Type:	Open Cluster
Location:	★
Rating:	★ ★
Best Seen:	Winter

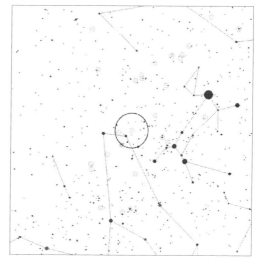

Map courtesy *Mobile Observatory*

This open star cluster is quite easily found, just to the west of Xi Puppis. It's a medium-sized cluster, quite dense with the vast majority of stars being blue-white with many other fainter stars visible with averted vision.

To get the best view, you'll want to get away from the city as I could barely see it at 26x. At that magnification, it appeared small but I was still able to see two bright stars at the tip of a group of much fainter stars. The group looked like a curved triangle.

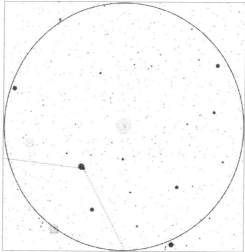

Finderscope view courtesy *Mobile Observatory*

The view from darkened skies is much better as many of the fainter stars are more easily visible. At 35x, the cluster made me think of a butterfly with the two bright, orange stars on the tip of the western wing.

Whether you're observing from the suburbs, the country or from the city itself, the cluster stands up well to higher magnification. I found the best view to be had at around 75x. Increasing the magnification to about 100x only made the cluster appear larger, rather than richer. (A larger telescope, with greater light gathering power, would almost certainly reveal more stars.)

The cluster was first recorded by Charles Messier in March 1781, just a week after William Herschel discovered the planet Uranus. The cluster lies about 3,600 light years away and is thought to contain about 80 stars and is about 25 light years in diameter.

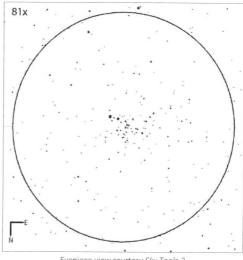

Eyepiece view courtesy *Sky Tools 3*

The Praesepe

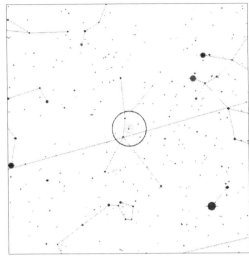
Map courtesy *Mobile Observatory*

Designation(s):	Messier 44
Constellation:	Cancer
R.A.:	08h 40m 22s
Declination:	+19° 40' 19"
Object Type:	Open Cluster
Location:	★★
Rating:	★★★
Best Seen:	Winter and Spring

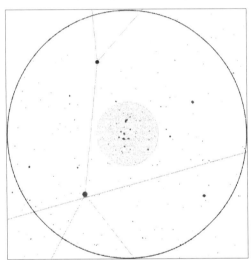
Finderscope view courtesy *Mobile Observatory*

Known since antiquity, the traditional name for the cluster is the Praesepe, which means "the manger." Another popular name is the Beehive Cluster. Personally, it always gives me the impression of a swarm of bees, but I also like the traditional name.

Whatever name you choose, you'll find spotting the cluster to be a good test of your environment. In theory, you should be able to see it with just your eyes but finding it can be a little problematic as it's actually brighter than any of the stars that form the constellation of Cancer itself.

Look midway between the mid-section of Gemini and Regulus, the brightest star in Leo, the lion. With luck, you should be able to see a tiny, misty patch. I've seen it from the suburbs but in the city I've always needed a pair of binoculars to help me.

It's a large cluster and, consequently, one that's best suited to lower magnifications. Even at 35x it'll barely fit within the field of view. But it's a very nice view, all the same with dozens of stars glinting against the night. The majority of stars appear blue-white, indicating a young age. Several others are older orange giants.

There are a lot of doubles to be found here, but look toward the center: can you see a mini Cepheus? It always leaps right out at me!

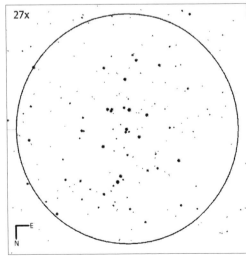
Eyepiece view courtesy *Sky Tools 3*

Iota Cancri

Designation(s):	Iota Cancri
Constellation:	Cancer
R.A.:	08h 46m 42s
Declination:	+28° 45' 36"
Object Type:	Multiple Star
Location:	★ ★ ★
Rating:	★ ★ ★
Best Seen:	Winter and Spring

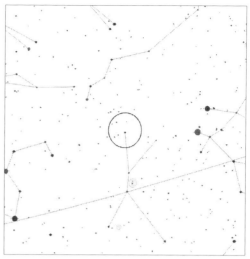

Map courtesy *Mobile Observatory*

Iota Cancri is like a pale Albireo for late winter and early spring. Unfortunately, it's not a prominent star, which means you might have difficulty finding it from the city. Scan the area about midway between Pollux and the top of the backwards question-mark that represents Leo's head.

Through a finderscope or binoculars, you'll see 53 and Rho Cancri - another, wider coppery double pair to the east but within the same field of view. This is a nice sight for binoculars and may be a consolation prize for anyone observing without a telescope.

Once you've found Iota, you'll find it easy to split with a low powered eyepiece. Just 26x will reveal both components. The primary is pale gold and appears to be about two or three times brighter than the very pale blue secondary.

Finderscope view courtesy *Mobile Observatory*

Although it looks good at all magnifications, it's probably best at about 100x where a decent distance appears between the pair. As a bonus, I've noted that the secondary shows definite signs of purple and deep blue when observed at higher powers.

The pair is not a true multiple star system – it's what's known as an *optical* binary. In other words, the stars only appear close to another due to chance alignment.

Eyepiece view courtesy *Sky Tools 3*

Messier 67

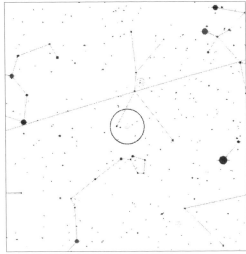
Map courtesy *Mobile Observatory*

Designation(s):	Messier 67
Constellation:	Cancer
R.A.:	08h 51m 20s
Declination:	+11° 48' 43"
Object Type:	Open Cluster
Location:	★
Rating:	★ ★
Best Seen:	Winter and Spring

Finderscope view courtesy *Mobile Observatory*

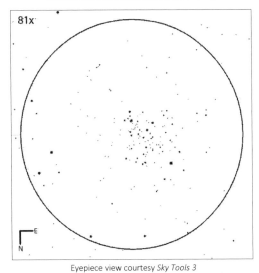
Eyepiece view courtesy *Sky Tools 3*

This late Winter / early Spring cluster is reminiscent of Messier 37 in Auriga. It's a tightly compacted cluster of faint blue-white stars that mostly appear to be about the same magnitude. There are a couple of notable exceptions, however.

It's barely visible from the city but can be glimpsed at 26x and using averted vision will help to bring out the faintest stars. In particular, look for a bright orange star that appears on the edge of the shield-shaped cluster. You should also see a pair of faint stars on the opposite edge but the remaining stars might be difficult.

Increasing the magnification will help – it will still fit within the field of view at about 90x - but the best views are reserved for those observing from a dark location.

Even at 35x, you'll see a glimmering diamond shape that appears to contain hundreds of tiny stars. To me, again, like Messier 37, it gave me the impression of sugar or even diamond dust on velvet but besides the overall shape of the cluster itself I didn't notice any particular patterns.

At more than three billion years old the cluster is one of the oldest known and lies at a distance of over 2,500 light years. Containing more than five hundred stars, it's thought to be about ten light years in diameter.

Regulus

Designation(s):	Alpha Leonis
Constellation:	Leo
R.A.:	10h 08m 22s
Declination:	+11° 58' 02"
Object Type:	Multiple Star
Location:	★ ★ ★
Rating:	★
Best Seen:	Spring

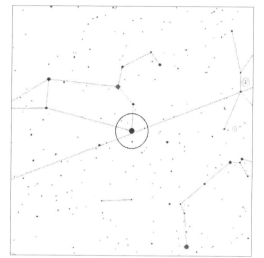

Map courtesy *Mobile Observatory*

Regulus is one of the brightest stars in the sky and is also one of the most overlooked – at least by experienced amateurs.

Through binoculars or a finderscope, you'll see the reasonably bright stars Nu and 31 Leonis within the same field of view, but look carefully with binoculars and you might see something else.

You'll need to hold your binoculars steady and you may have to wait for the air to be still but you might just see a very faint companion to the north-west. (If your finder magnifies the view, try your luck with that too.)

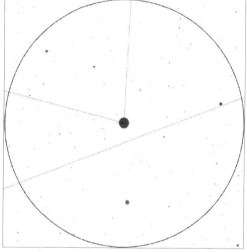

Finderscope view courtesy *Mobile Observatory*

The pair are much more apparent at low power in a small telescope. At 27x, you'll see Regulus as a brilliant white star, much brighter than the bluish secondary. It'll be an easy and wide split but if you want the secondary to be more easily seen, you'll need to up the power some more. I found that 50x provided quite a nice view.

Regulus is the brightest star in the constellation of Leo the Lion and, at 79 light years away, is relatively close to Earth. The pair are a part of a true multiple star system with both components orbiting one another once every two million years.

Each component is itself double, making this a four star system in all.

Eyepiece view courtesy *Sky Tools 3*

Adhafera

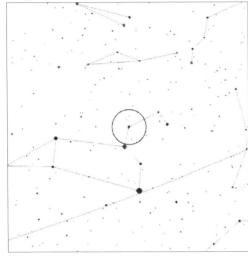

Map courtesy *Mobile Observatory*

Designation(s):	Zeta Leonis
Constellation:	Leo
R.A.:	10h 16m 41s
Declination:	+23° 25' 02"
Object Type:	Multiple Star
Location:	★ ★ ★
Rating:	★ ★
Best Seen:	Spring

Finderscope view courtesy *Mobile Observatory*

Eyepiece view courtesy *Sky Tools 3*

Adhafera is another overlooked star in Leo, the Lion. Through the finder or binoculars you'll see a fainter star, 35 Leonis, to the south. This star is not associated with Adhafera and its position is purely due to chance.

Specifically, Adhafera is about 260 light years away while 35 Leonis is about 100 and both stars are moving in different directions.

Unfortunately, I don't have any notes from binocular observations, but you may also be able to spot another, fainter companion to the north-east. It'll appear very close to Adhafera itself with the pair best observed through a small telescope.

At 27x you'll see a wide pair of stars comprising of a lemony-white primary that's about 1½ times brighter than its white companion.

As with many double stars, I've noted some variance in the colors; sometimes the primary appears to be purely white while the secondary can appear peachy.

Adhafera is notable for another important reason – the Leonid meteors appear to radiate from a point just to the west of the star. This shower peaks around November 18th and usually produces about ten or twenty shooting stars per hour. However, on occasion, the shower has been known to exceed a hundred.

Algieba

Designation(s):	Gamma Leonis
Constellation:	Leo
R.A.:	10h 19m 58s
Declination:	+19° 50' 30"
Object Type:	Multiple Star
Location:	★ ★ ★
Rating:	★ ★ ★
Best Seen:	Spring

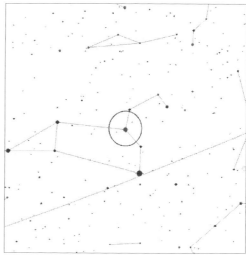
Map courtesy *Mobile Observatory*

Algieba is an easy, wide double star that can be easily seen with either binoculars or even your own unaided eyes. The primary appears gold and about two or three times as bright as the blue-ish companion. You might also notice a third, slightly fainter star that forms a triangle with the two brighter stars.

The real magic happens when you turn a telescope toward the trio. You won't notice anything different until you get to about 50x or higher. The gold star may be barely split at this magnification with a pale white gold companion appearing close beside it.

If you're having difficulty with the pair, it might be due to your location and/or your observing conditions. I've been able to comfortably split it at under 100x from the suburbs but from the city it's barely split at even 162x. (This, no doubt, is also due to the higher humidity of my location and the fact that I was observing the star in June, when temperatures were still 70°F at night.)

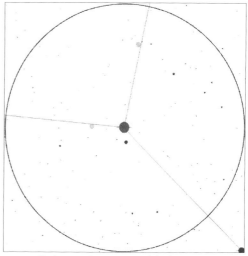
Finderscope view courtesy *Mobile Observatory*

What's the lowest magnification you can use to split the star and see the secondary? Find a magnification where the secondary is clearly seen and then trying working your way back down until it disappears again. At what point does Algieba go from being a clean split to elongated and then just a single point of light?

Eyepiece view courtesy *Sky Tools 3*

Denebola

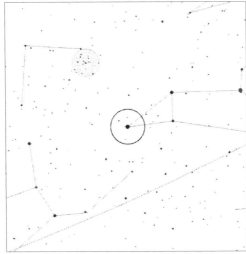

Map courtesy *Mobile Observatory*

Designation(s):	Beta Leonis
Constellation:	Leo
R.A.:	11h 49m 04s
Declination:	+14° 34' 19"
Object Type:	Multiple Star
Location:	★ ★ ★
Rating:	★ ★
Best Seen:	Spring

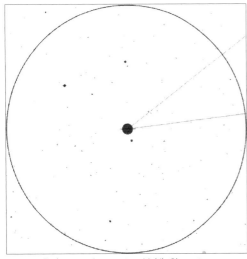

Finderscope view courtesy *Mobile Observatory*

Eyepiece view courtesy *Sky Tools 3*

Denebola is a bright star that marks the tail of Leo the Lion – and it's not one you usually find highlighted in an astronomy book.

In fact, I really didn't pay any attention to it until I bought a small 70mm reflector for my son and was looking for easy objects for him to observe. In essence, it was the beginnings of the very book you're reading now and demonstrates the benefits of going off the beaten track and sometimes just randomly exploring the night sky.

Through binoculars or a finderscope, you'll see a wide pair of stars, similar to Algieba. Both are white with the primary about four times brighter than the secondary.

Similarly, if you observe the star through a small telescope – even at a low magnification of 15x – you'll easily see the pair but a third, much fainter star will become apparent. It's a lot fainter than the primary and can be seen about a quarter of the way between the first two stars. You may need to use averted vision to see it properly.

I then decided to take a look through my 130mm 'scope and increased the magnification even further. Once I got to 108x, I got another surprise. While all three stars could still be seen within the field of view, a fourth star now appeared, fainter than the third. Look for it about midway between the first two and a little out of line with the others.

Algorab

Designation(s):	Delta Corvi
Constellation:	Corvus
R.A.:	12h 29m 52s
Declination:	-16° 30' 56"
Object Type:	Multiple Star
Location:	★ ★ ★
Rating:	★ ★
Best Seen:	Spring

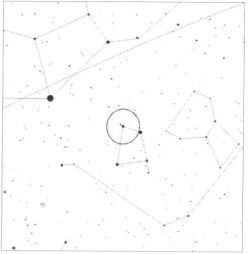

Map courtesy *Mobile Observatory*

Algorab is the third brightest star in the constellation of Corvus the Crow. In Greek mythology, the god Apollo sent the crow to fetch him a cup of water but the bird stopped to eat figs and was slow to return. To punish the bird, Apollo threw him and the cup into the sky. Consequently, Crater, the Cup, appears to the west of Corvus and will forever be out of reach of the bird.

Through a finder or binoculars, you'll see a wide, unrelated companion star, just to the north-west. You may even be able to see it with just your eyes. However, you'll need a telescope to see both components of Algorab itself.

Finderscope view courtesy *Mobile Observatory*

The star was usually split at 26x but it was dependent upon the quality of the atmosphere, especially from the city. It can be a clean split but the secondary is very faint and may not be seen unless you use a higher magnification first.

It was a lot easier under suburban skies at 35x. The primary appeared white with a very pale and faint rusty colored secondary close beside it. Increasing the magnification might improve the view but I've noticed the colors appear to fade in doing so.

In reality, the two stars are moving along a similar path through space but they're not thought to be a true multiple star system.

Eyepiece view courtesy *Sky Tools 3*

Cor Caroli

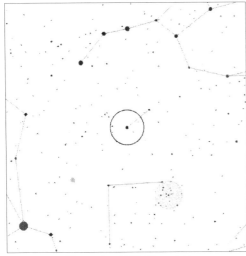
Map courtesy *Mobile Observatory*

Designation(s):	Alpha Canum Venaticorum
Constellation:	Canes Venatici
R.A.:	12h 56m 02s
Declination:	+38° 19' 06"
Object Type:	Multiple Star
Location:	★ ★ ★
Rating:	★ ★ ★
Best Seen:	Spring and Summer

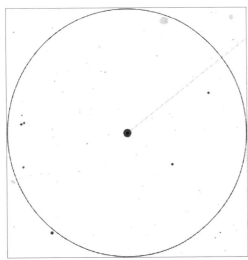
Finderscope view courtesy *Mobile Observatory*

Eyepiece view courtesy *Sky Tools 3*

The name Cor Caroli actually means "Charles's Heart" and it originates from the 17th century when it was said the star shone particularly brightly upon the King's return to England.

It's a showpiece multiple star for the Spring and Summer skies and another favorite with astronomers throughout the northern hemisphere. Even at a low magnification of just 26x the star is clearly split and provides a beautiful view. Increasing the power to about 50x or 60x will probably give you the best view.

The primary is white and appears about three times brighter than the secondary. However, I've noted slightly different colors for the secondary; more often than not, I've noted that it appears white too, but on occasion it seems to hold a creamy or pale gold color. (I've even noted a grey-greenish color at a magnification of 182x.)

What colors do you see?

(Most other sources describe the pair as both being either white or blue-white in color but some have also noted the different colors reported by observers.)

Cor Caroli is a true binary star system some 110 light years away with the pair orbiting one another roughly every 8,300 years. Curiously, the primary component is highly magnetic and has a magnetic field thousands of times stronger than the Earth's.

Mizar and Alcor

Designation(s):	Zeta and 80 Ursae Majoris
Constellation:	Ursa Major
R.A.:	13h 23m 56s
Declination:	+54° 55' 31"
Object Type:	Multiple Star
Location:	★ ★ ★
Rating:	★ ★ ★
Best Seen:	Spring and Summer

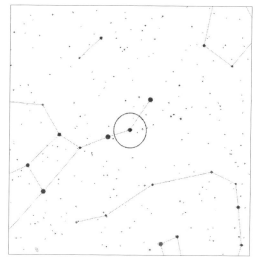

Map courtesy *Mobile Observatory*

Mizar and Alcor are a wide pair of stars that have been considered a test of eyesight since ancient times. As both stars are well within the limits of naked eye visibility, I've never had any trouble seeing them both – even from the city.

If you're familiar with the seven stars of the Dipper (or the Plough, as it's known in the United Kingdom) you'll find the pair in the middle of the curved line of three that make up the handle. Mizar is, of course, immediately visible but you might need to stare for a few moments to spot Alcor.

The pair make for an attractive sight in binoculars with Mizar being about twice as bright as Alcor and both stars appearing blue-white.

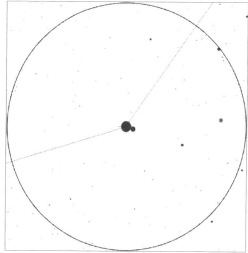

Finderscope view courtesy *Mobile Observatory*

However, like many other doubles, it's only when you turn a telescope toward the stars that everything comes into view. At a low power of just 27x, Mizar is clearly and easily split with the secondary also blue-white and about twice as faint as the primary.

While you're in the area, look out for a third star that forms a triangle with Alcor and the two components of Mizar.

Mizar and Alcor is not a true double star system but Mizar itself is a quadruple star some 85 light years away.

Eyepiece view courtesy *Sky Tools 3*

Arcturus

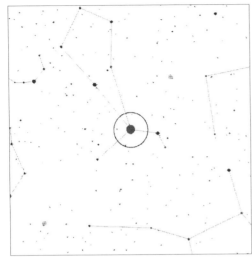
Map courtesy *Mobile Observatory*

Designation(s):	Alpha Boötis
Constellation:	Boötes
R.A.:	14h 15m 40s
Declination:	+19° 10' 56"
Object Type:	Multiple Star
Location:	★ ★ ★
Rating:	★
Best Seen:	Spring and Summer

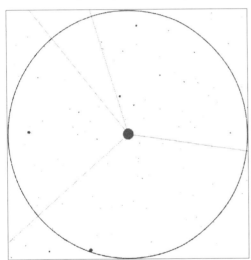
Finderscope view courtesy *Mobile Observatory*

Eyepiece view courtesy *Sky Tools 3*

Arcturus is the fourth brightest star in the sky and, at just under 37 light years away, one of the closest to the Earth.

It's a red giant star that appears orange to the unaided eye but despite this, at 27x I've noted that it shows a brilliant, white gold hue. Other observers have noted a range of colors including yellow, peach and topaz.

(If you want to see the star truly sparkle, wait until it's low on the horizon. Like Sirius, you'll find it flashing a multitude of colors when it's close to setting in the west.)

This has always been something of a favorite star for me. It's easily found by following the curved tail of Ursa Major, the Great Bear, down towards the south. ("Arc down to Arcturus.") It appears at the bottom of kite-shaped Boötes, a large constellation representing a herdsman of the same name. (The name Arcturus actually means "guardian of the bear.")

Arcturus is an old star, maybe seven billion years old, but is still over a hundred times more luminous than the Sun. As one of the fastest moving stars, it's currently drawing nearer and will be at its closest in about 4,000 years. In the far more distant future, it will eventually shed its outer layers and shrink down to a white dwarf, forming a planetary nebula in the process. Unfortunately, that sight will only be seen by our descendants as you and I will both be long gone!

Delta Boötis

Designation(s):	Delta Boötis
Constellation:	Boötes
R.A.:	15h 15m 30s
Declination:	+33° 18' 53"
Object Type:	Multiple Star
Location:	★ ★ ★
Rating:	★ ★ ★
Best Seen:	Spring and Summer

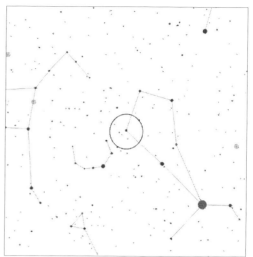

Map courtesy *Mobile Observatory*

There are a number of multiple stars to be found in Boötes but Delta Boötis is one of the easiest to find. It's a relatively bright star that marks the upper-left corner of the kite-shaped constellation.

Although a fairly wide double, I haven't been able to split the pair with binoculars but others have noted that it's been done so it's definitely worth a try. (I was using a smaller pair of 8x30's so a standard pair of 10x50's might be just enough to split them.)

If you're using a small telescope, a low magnification of about 25x should easily do the trick.

At that magnification, it appears to be a wide and pretty pair with a nice color contrast between the two stars. The pale yellow-white primary appears about five times brighter than pale blue secondary. Add to that a number of other stars that appear within the same field of view and you have a sight worth looking out for.

Delta is thought to be a true double star system, lying some 121 light years from Earth. The primary is a giant star nearing the end of its life and some ten times larger than our own Sun while the secondary is a dwarf star a little smaller than the Sun. The pair are thought to orbit one another once every 120,000 years.

Finderscope view courtesy *Mobile Observatory*

Eyepiece view courtesy *Sky Tools 3*

Zuben Elgenubi

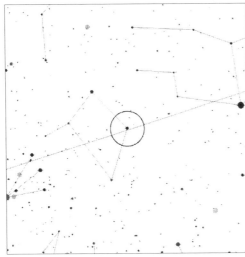
Map courtesy *Mobile Observatory*

Designation(s):	Alpha Librae
Constellation:	Libra
R.A.:	14h 50m 41s
Declination:	-15° 59' 50"
Object Type:	Multiple Star
Location:	★ ★ ★
Rating:	★ ★
Best Seen:	Spring

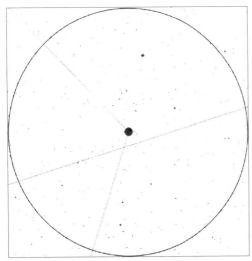
Finderscope view courtesy *Mobile Observatory*

Eyepiece view courtesy *Sky Tools 3*

Zuben Elgenubi – or Alpha Librae, if you prefer – is a wide pair of stars easily seen with both binoculars and telescopes at low power. In fact, this is one example where the object is best at low power as increasing the magnification too much reduces the aesthetic appeal of the stars.

Both stars are white but the primary is about 1½ times brighter than the secondary.

The name Zuben Elgenubi is Arabic and means "the southern claw." In case you're wondering how that relates to a constellation that represents scales, it originates from a time when the stars of Libra were still considered part of nearby Scorpius, the Scorpion.

Despite it having the designation of Alpha Librae, it is in fact the second brightest star in the constellation. It's thought to be a true binary system (as opposed to being just a chance alignment of two stars) that lies about 77 light years away. The pair are estimated to obit one another once every 200,000 years.

Each star is itself a binary system, making four components in all with a suspected fifth member – KU Librae – appearing nearby.

Graffias

Designation(s):	Beta Scorpii
Constellation:	Scorpius
R.A.:	16h 05m 26s
Declination:	-19° 48' 20"
Object Type:	Multiple Star
Location:	★ ★ ★
Rating:	★ ★ ★
Best Seen:	Spring and Summer

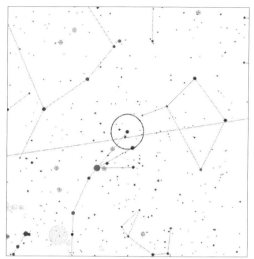
Map courtesy *Mobile Observatory*

Graffias is an interesting star – and also slightly confusing as the name applies to both Beta and Xi Scorpii. To clarify, Beta appears in the center of the finderscope depiction to the left and is the star we're interested in.

If you're using binoculars, you should easily see Omega^1 and Omega^2, a fairly wide pair of white stars nearby. These are the two stars that can be seen off center, just to the lower left of Beta in the finderscope view to the left. You might also notice a couple of other fainter stars in the field of view.

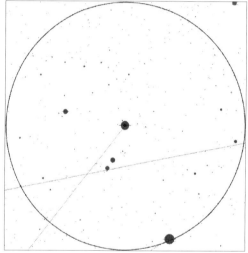
Finderscope view courtesy *Mobile Observatory*

This in itself is a nice enough view, but if you turn your telescope toward Beta you should be able to split it at 27x without any trouble.

Both components of the primary appear white (or sometimes blue-white) with the secondary being about three times fainter than the primary. It's a pretty nice view, even at this magnification.

Look carefully at Beta's secondary as I've noticed it showed signs of a violet color at 81x. However, this was while observing from the city, so it's possible the atmospherics were not great at the time.

There are, in all, six components to Beta Scorpii and the entire system lies some 400 light years away.

Eyepiece view courtesy *Sky Tools 3*

Jabbah

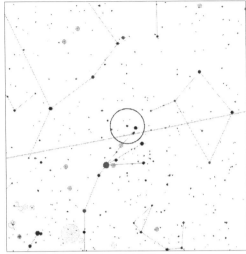
Map courtesy *Mobile Observatory*

Designation(s):	Nu Scorpii
Constellation:	Scorpius
R.A.:	16h 12m 00s
Declination:	-19° 27' 39"
Object Type:	Multiple Star
Location:	★★
Rating:	★★★
Best Seen:	Spring and Summer

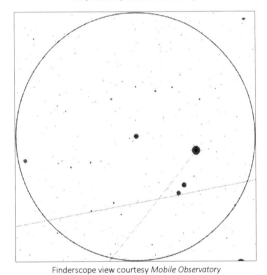
Finderscope view courtesy *Mobile Observatory*

If you're able to locate Graffias, you should have no difficulty spotting Jabbah as it lies to the east and within the same finderscope field of view.

Unfortunately, despite what you might think (or hope) Jabbah has nothing to do with *Star Wars* - but it's still a pretty interesting sight for telescope observers at both low and medium magnification.

At about 30x you'll see a pair of white stars of similar brightness at one corner of a triangle of three. But increase the magnification and what do you see?

At 81x one of the stars appeared blue-ish, with hints of violet but increasing the magnification again to 107x reveals a third companion.

Look carefully at the fainter component and you might see that it's split into two white stars of almost equal magnitude.

Like many other multiple stars, there are other components to this system that can't be seen with amateur equipment. In fact, it's thought there may be up to seven stars in this system, which would make for an amazing (and possibly blinding) sight if you could ever get there. And you thought a twin sunset over the deserts of Tatooine was cool!

Eyepiece view courtesy *Sky Tools 3*

Messier 80

Designation(s):	Messier 80
Constellation:	Scorpius
R.A.:	16h 17m 02s
Declination:	-22° 58' 34"
Object Type:	Globular Cluster
Location:	★
Rating:	★
Best Seen:	Spring and Summer

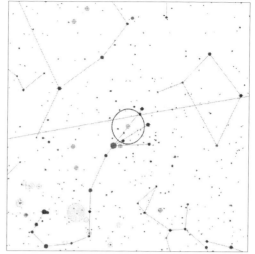

Map courtesy *Mobile Observatory*

Messier 80 is the *other* globular cluster that's easily found within the constellation of Scorpius, the Scorpion. It's often overlooked in favor of it's bigger and brighter neighbor, Messier 4, but that's not to say it's not worth seeking out.

From the suburbs, I was able to easily find the cluster at 35x. It appeared as a grey smudge close to a brighter star to the west and a group of stars to the east.

Although it's smaller than Messier 4, it appeared a little brighter to me with a bright core. Unfortunately, at both 70x and 182x there didn't seem to be much change and I wasn't able to resolve the cluster into individual stars.

It was another ten years before I tried again, but this time it was from the brighter skies of the city and I had less success than before. I was able to spot it at 27x, when it appeared as a very small, faint, circular grey sphere close to a star, but again, I wasn't able to improve the view.

Finderscope view courtesy *Mobile Observatory*

So why hunt it down at all? Well, think about what you're seeing here. Messier 80 is over 32,000 light years away and is thought to contain about 200,000 stars, all crammed into an area some 95 light years across. In comparison, there are only about 500 stars within 100 light years of the Earth, which means it contains about 400 times as many stars as our local stellar neighborhood.

Are you still a little disappointed?

Eyepiece view courtesy *Sky Tools 3*

Messier 4

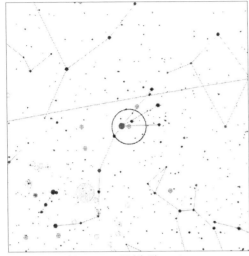
Map courtesy *Mobile Observatory*

Designation(s):	Messier 4
Constellation:	Scorpius
R.A.:	16h 23m 35s
Declination:	-26° 31' 33"
Object Type:	Globular Cluster
Location:	★
Rating:	★★
Best Seen:	Spring and Summer

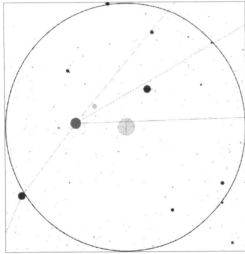
Finderscope view courtesy *Mobile Observatory*

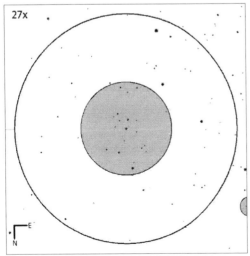
Eyepiece view courtesy *Sky Tools 3*

If Messier 80 failed to impress, try your hand at Messier 4. It appears very close to Antares, the reddish star that marks the beating heart of Scorpius the Scorpion.

In fact, it's fair to say the cluster is almost unmissable – unless you're trying to find it with binoculars in the city. I've never had any luck there, but if you live under dark skies you shouldn't have much problem at all.

As with almost every deep sky object (such as star clusters, nebulae and galaxies) you'll see more under darker skies and Messier 4 is no exception.

I've been able to pick it out at low power from both the suburbs and the city, but a magnification of about 30x is only enough to show it as a very faint and fuzzy sphere.

At around 50x the cluster appeared elongated, with a band of bright stars stretching across the center and some resolution around the edges with averted vision.

City dwellers may not be so lucky but I've noted some resolution at 81x – again, with averted vision.

In comparison, at around 80x or 90x the view from suburban skies improves quite considerably and the cluster appears quite spectacular. I've noted that thousands of stars could be seen with averted vision and the cluster looked like sugar or salt spilt on black satin.

The Butterfly Cluster

Designation(s):	Messier 6
Constellation:	Scorpius
R.A.:	17h 40m 21s
Declination:	-32° 15' 15"
Object Type:	Open Cluster
Location:	★ ★
Rating:	★ ★ ★
Best Seen:	Summer

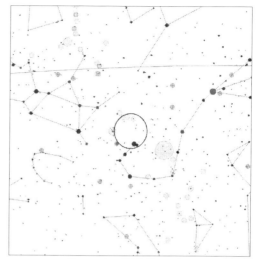
Map courtesy *Mobile Observatory*

The Butterfly cluster is one of two (the other being Messier 7) that won't be visible from the United Kingdom and skims the horizon from North America. If you live in the United States, you shouldn't have much problem as long as the whole of Scorpius can be seen from your location.

To find it, place the two stars representing the sting of the Scorpion's tail on the edge of your field of view in about the five o'clock position. (Messier 7 may be just outside the view at about the nine o'clock position.) The Butterfly cluster may appear near to the eleven o'clock position. I've noted that it's barely visible from suburban skies and is the middle of the three circles in the finderscope depiction.

Through binoculars it's about half the size of Messier 7 and looks like tiny, very fine granules of star dust in a triangular shape that points toward the east.

Finderscope view courtesy *Mobile Observatory*

Denser and more compact than Messier 7, the cluster provides a beautiful view at 35x with its butterfly shape being quite apparent. The brightest five stars form a shape like a flattened Pleiades. Four of those are blue-white with the fifth, on the south-western side, being white, possibly orange-ish.

The majority of the stars appear much fainter, look like diamond dust and must number in the hundreds.

Eyepiece view courtesy *Sky Tools 3*

Messier 7

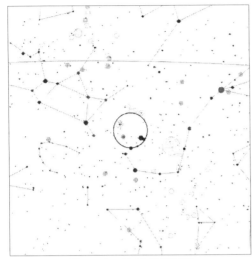

Map courtesy *Mobile Observatory*

Designation(s):	Messier 7
Constellation:	Scorpius
R.A.:	17h 53m 51s
Declination:	-34° 47' 34"
Object Type:	Open Cluster
Location:	★ ★
Rating:	★ ★
Best Seen:	Summer

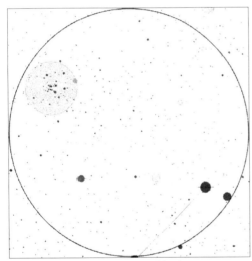

Finderscope view courtesy *Mobile Observatory*

Eyepiece view courtesy *Sky Tools 3*

Close to the Butterfly Cluster in the sky, Messier 7 is larger, brighter and is consequently a little easier to see. You may be able to spot it with just your eyes under a clear, dark sky.

With the two stars of the scorpion's stinger at roughly the four o'clock position in your finder, it should be quite apparent on the opposite edge of the field of view. In fact, the two stars will point you right to it.

A low power magnification is all you'll need to enjoy this cluster. It's not as dense as the Butterfly and, in my opinion, is not as attractive. At 35x I've noted plenty of sparsely scattered stars that appear to form the shape of a bow tie (bow ties are cool, by the way.)

You may also see some gold stars among the many blue-white members; in particular, look out for an apparent double at the very heart of the cluster. Both stars are of equal brightness but with one being pale gold and the other blue-white.

Increasing the magnification to 54x may also reveal a small trio of stars forming an equilateral triangle on the eastern edge.

Messier 7 lies about 980 light years away and is thought to be about 25 light years in diameter.

The Keystone Cluster

Designation(s):	Messier 13
Constellation:	Hercules
R.A.:	16h 41m 41s
Declination:	+36° 27' 37"
Object Type:	Globular Cluster
Location:	★★
Rating:	★★★
Best Seen:	Spring and Summer

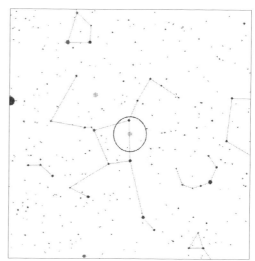
Map courtesy *Mobile Observatory*

The Keystone Cluster is by far the best globular cluster visible from the northern hemisphere.

If you have sharp eyesight and you live under clear, dark skies, you may be able to spot the cluster with just your eyes, but (more likely) you'll probably need binoculars to see it.

Through my small 8x30's from suburbia it appeared very small and looked like a faint, fuzzy star close to two other stars. It had a star-like core and appeared comet-like.

The darker your sky, the more you'll see and you can certainly reveal a lot of detail under the right conditions. Unfortunately, it's a disappointment from the city and a shadow of its true self. Just 27x will allow you to see the cluster but it's not much better than using binoculars and increasing the magnification doesn't improve the view.

From the suburbs it's another story. At 35x I noted the bright core extending some way toward the edge, where it then faded rapidly. I also noted some resolution and – best of all – chains of stars extending out from the core.

I've found a magnification of about 100x to be best for small 'scopes. I've often thought it like some strange sea creature, with tentacles reaching out from the depths of space.

Finderscope view courtesy *Mobile Observatory*

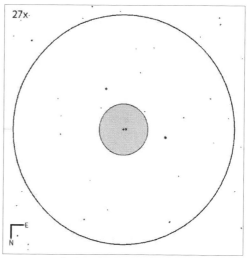
Eyepiece view courtesy *Sky Tools 3*

Rho Herculis

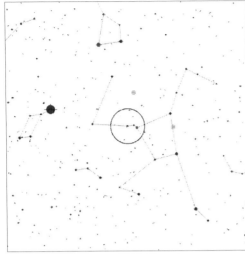
Map courtesy *Mobile Observatory*

Designation(s):	Rho Herculis
Constellation:	Hercules
R.A.:	17h 24m 12s
Declination:	+37° 08' 16"
Object Type:	Multiple Star
Location:	★ ★
Rating:	★ ★
Best Seen:	Spring and Summer

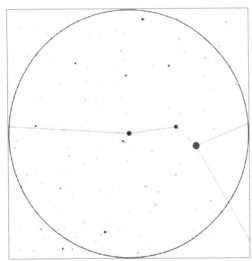
Finderscope view courtesy *Mobile Observatory*

Eyepiece view courtesy *Sky Tools 3*

Rho Herculi is a white star with a nearby blue companion that's easily seen with binoculars. You might also see a fainter, third companion close to that secondary.

A telescope won't reveal much else until you get to about 50x when the primary can be barely split. Both components appear white with the primary being slightly brighter than the secondary.

Increasing the power helps to reveal some color in the pair. A magnification of about 100x shows a clear split with the primary being a pale gold. The secondary has shown some interesting colors – I've noted olive green with flecks of mustard at 107x.

The system lies just over 400 light years away and comprises of a subgiant primary and a dwarf secondary. The system is thought to be about 300 million years old with a third, very faint suspected companion appearing some distance away.

Rho marks the shoulder of Hercules, the Hero. In ancient mythology, he was known as Heracles and was the mortal son of Zeus. He's famous for performing twelve labors, by order of King Eurystheus; his first was to slay a lion and his second was to slay a sea serpent. Both these creatures can now be found toward the west in the night sky, some way from Hercules, in the form of the constellations Leo and Hydra.

Kuma

Designation(s):	Nu Draconis
Constellation:	Draco
R.A.:	17h 32m 11s
Declination:	+55° 11' 03"
Object Type:	Multiple Star
Location:	★ ★ ★
Rating:	★ ★ ★
Best Seen:	Spring and Summer

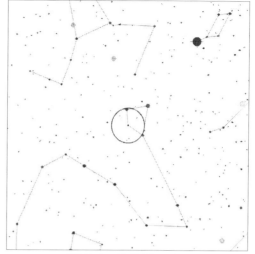

Map courtesy *Mobile Observatory*

Kuma is a gem and a very easy target for small telescopes in the northern hemisphere. I was able to split the pair with both my finderscope and my small 8x30 binoculars (although I had to hold them steadily to do it. Leaning against a wall will definitely help!)

This is a double that's best observed at low power. It's an easy and clean split at just 27x with both stars appearing to be brilliant white and of equal brightness. They've always made me think of a pair of eyes staring back at me from space – in fact, some observers have taken to calling them "the eyes of the dragon" for that reason.

A true multiple star system, the pair lie just under a hundred light years away with one component, Nu Draconis[2], having its own very close companion.

Finderscope view courtesy *Mobile Observatory*

The constellation of Draco, the Dragon, snakes around Polaris, the Pole Star, and is actually circumpolar. This means the constellation never sets when seen from most of the northern hemisphere but appears to spin about the pole star instead.

Its head appears close to Hercules, the Hero and appears to chase him about the pole. It's at its highest in the Summer when the constellation can be seen snaking over much of the northern horizon.

Eyepiece view courtesy *Sky Tools 3*

Messier 22

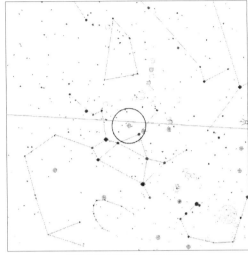
Map courtesy *Mobile Observatory*

Designation(s):	Messier 22
Constellation:	Sagittarius
R.A.:	18h 36m 24s
Declination:	-23° 54' 17"
Object Type:	Globular Cluster
Location:	★ ★
Rating:	★ ★
Best Seen:	Summer

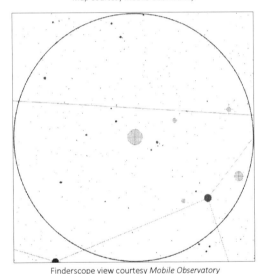
Finderscope view courtesy *Mobile Observatory*

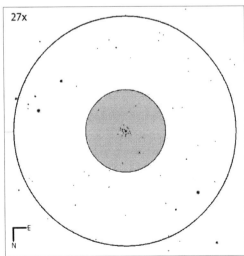
Eyepiece view courtesy *Sky Tools 3*

Messier 22, like the Keystone Cluster, is one of the best globulars in the entire night sky but unlike its cousin in Hercules, it's not well seen from the northern hemisphere.

This is a shame as it would certainly would be much better known if it were more easily seen. However, if you live in the United States it shouldn't be too much of a problem and even observers in the United Kingdom and Canada may still stand a chance.

It's conveniently located close to Kaus Borealis, the star that marks the top of the famous teapot of Sagittarius. It's easily seen with binoculars from suburban skies and appears as a small, nebulous sphere with no bright center and a uniform grey color.

A low magnification of 35x won't improve the view too much; it appears circular and could be easily mistaken for a nebula.

Increasing the magnification to 50x or higher and using averted vision might help reveal some resolution but take a look at its shape; does it still appear circular or has its appearance altered slightly? Does it look a little elongated to you?

Messier 22 is only 10,600 light years away (quite close for a globular) and contains about 70,000 stars.

The Double Double

Designation(s): Epsilon Lyrae
Constellation: Lyra
R.A.: 18h 44m 20s
Declination: +39° 40' 12"
Object Type: Multiple Star
Location: ★ ★ ★
Rating: ★ ★ ★
Best Seen: Summer

Map courtesy *Mobile Observatory*

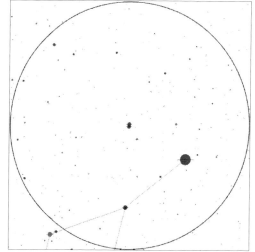

Finderscope view courtesy *Mobile Observatory*

Eyepiece view courtesy *Sky Tools 3*

Epsilon Lyrae, commonly known as the "Double Double," is one of the most famous and popular multiple stars in the sky – and for good reason.

First of all, turn a pair of binoculars toward it – what do you see? You might need to steady yourself against a wall, but you should easily be able to split the star into two close white components of equal brightness. It's a nice view with both Vega and Zeta Lyrae appearing within the same field of view.

In a small 'scope at low power (say, about 30x) you might notice something odd. I've seen both stars appear elongated, as though they were almost but not quite split.

Try increasing the magnification – first to about 50x, then about 75x. Are the stars split yet? Try again at about 100x and you should be able to barely split both stars into their components and the view is quite fascinating. Two pairs of bright white stars, with all four being about the same brightness. One pair might appear slightly brighter and wider than the other.

You're looking at a fascinating multiple star system some 162 light years away. One pair orbit each other once every 1,800 years while the other pair circle one another every 700 years. While quadruple stars are not unheard of, what's truly astonishing is that there may be as many as ten stars in this system!

Sheliak

Map courtesy *Mobile Observatory*

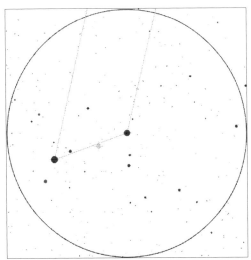
Finderscope view courtesy *Mobile Observatory*

Eyepiece view courtesy *Sky Tools 3*

Designation(s):	Beta Lyrae
Constellation:	Lyra
R.A.:	18h 50m 05s
Declination:	+33° 21' 46"
Object Type:	Multiple Star
Location:	★★★
Rating:	★★
Best Seen:	Summer

Sheliak is a relatively wide double just beyond the range of regular binoculars. However, a small 'scope shouldn't have any problem resolving the pair at low magnification. The primary is pure white and about five or six times brighter than the blue companion.

Look carefully and you'll see the pair just off-center of a small triangle of faint stars. Increasing the magnification to around 100x will definitely make this easier to see.

Sheliak is one of those stars that makes you wish interstellar travel were possible. If you could visit this star, you'd see something quite amazing.

It lies about 950 light years away and its primary component is actually a very close double. The smaller of the two was once the giant star of the system but as it grew in size, its companion began to pull matter away from it.

In other words, the larger star was cannibalized by its smaller companion until the roles were reversed. The companion is now the dominant member of the pair and is surrounded by a disk of matter stripped away from the former giant. The two stars now orbit one another almost every thirteen days causing the star to dim slightly as seen from Earth. Try comparing its brightness to nearby stars and you may notice it fade and brighten over the intervening nights.

The Ring Nebula

Designation(s): Messier 57

Constellation: Lyra

R.A.: 18h 53m 35s

Declination: +33° 01' 45"

Object Type: Planetary Nebula

Location: ★

Rating: ★ ★ ★

Best Seen: Summer

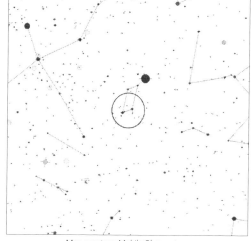

Map courtesy *Mobile Observatory*

A favorite of mine ever since I first saw it, it's very appropriately named because it truly looks like a smoke ring in space.

Surprisingly, I've been able to see it from both the city and also the suburbs. At 27x it appeared circular, dark grey, small and clearly not a star. However, you won't be able to see it as a ring until you increase the magnification.

At 70x I was barely able to see the central hole and the ring-shape, but I had to use averted vision to spot it. The shape becomes more apparent as you continue to increase the magnification but you'll need to always use averted vision to see the central hole.

At 161x I noted that it still appeared circular and cloud-like but that the nebula was well defined and the hole very apparent with averted vision. Focusing proved to be difficult while looking directly at it.

What you're seeing here is a dying star, some 2,000 light years away. The Ring Nebula is probably the best example of a planetary nebula – so-called because they look like planets when observed telescopically. It's actually a shell of gas thrown off from a giant star as it shrinks down to a white dwarf. You can actually see this star, in the center of the nebula, but you'll need a larger telescope to spot it.

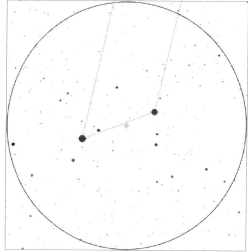

Finderscope view courtesy *Mobile Observatory*

Eyepiece view courtesy *Sky Tools 3*

The Wild Duck Cluster

Map courtesy *Mobile Observatory*

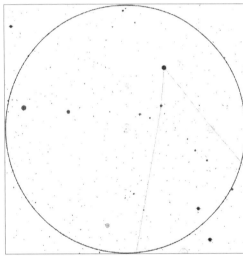
Finderscope view courtesy *Mobile Observatory*

Eyepiece view courtesy *Sky Tools 3*

Designation(s):	Messier 11
Constellation:	Scutum
R.A.:	18h 51m 06s
Declination:	-06° 16' 12"
Object Type:	Open Cluster
Location:	★★
Rating:	★★
Best Seen:	Summer

The Wild Duck Cluster is a small open star cluster found just to the south-west of Aquila, the Eagle. It can be easily seen with binoculars and can be found by following a slightly curved line of stars from Delta Aquilae, at the bottom of the constellation. (Alternatively, if you can easily locate Beta Scutum in your finderscope you should be able to spot the cluster within the same field of view.)

The cluster is disappointing from the city as it's small, faint and difficult to see at low power. You'll definitely need to use averted vision to really see anything of it.

If, however, you live away from the city lights, you're in for a treat. Through binoculars it appears small, compact, faint and globular – cloud-like, in fact – with a star-like point on the eastern edge.

Through a small 'scope at 35x you'll see a small, granular V shape with a single bright star at the tip. It's not hard to see how the cluster got its name as it looks like ducks flying south for the winter. You'll also see a few other bright stars appear within the same field of view, most notably a nearby double pair.

It's one of those clusters that definitely improves as you increase the magnification. Many more stars appear at just 54x with so many stars becoming apparent that it starts to lose its shape. The stars themselves are blue and are almost all of a uniform brightness

The Coathanger Cluster

Designation(s):	Collinder 399
Constellation:	Vulpecula
R.A.:	19h 26m 05s
Declination:	+20° 13' 14"
Object Type:	Asterism
Location:	★★
Rating:	★★
Best Seen:	Summer

Map courtesy *Mobile Observatory*

Finderscope view courtesy *Mobile Observatory*

Eyepiece view courtesy *Sky Tools 3*

The Coathanger Cluster is a large group of stars that's actually best observed with a very low power. In fact, you may prefer to observe it with either a pair of binoculars or even your telescope's finder.

To find it, you'll first need to locate the constellation of Sagitta, the Arrow, located a little under halfway between Albireo in Cygnus and Altair in Aquila. Look for the two stars that mark the feathers and place them at about the seven o'clock position in your field of view. The cluster should then be around the two o'clock mark.

The shape of the cluster should be quite apparent (albeit upside-down) through the finder or binoculars. When you turn your telescope toward it, make sure you're using your lowest power eyepiece first as you'll have difficulty fitting it into the field of view at even 35x.

The coathanger is formed by a chance alignment of eight or nine stars with the brightest forming the hook of the hanger.

The group has been known since antiquity, having first been recorded by the Persian astronomer al-Sufi around the year 964. The cluster also goes by the name Brocchi's Cluster, named for D. F. Brocchi, a stellar cartographer who mapped the cluster in the 1920's.

The Dumbbell Nebula

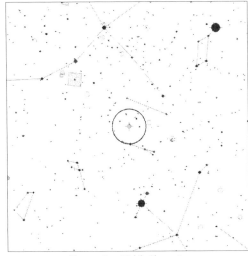
Map courtesy *Mobile Observatory*

Designation(s):	Messier 27
Constellation:	Vulpecula
R.A.:	19h 59m 36s
Declination:	+22° 43' 16"
Object Type:	Planetary Nebula
Location:	★
Rating:	★ ★
Best Seen:	Summer

Finderscope view courtesy *Mobile Observatory*

The Dumbbell Nebula, although faint, is actually relatively easy to find once you know where to look. You'll need to first find the constellation of Sagitta, the Arrow, about midway between Altair in Aquila and Albireo in Cygnus.

Look for Gamma Sagittae, one of the brighter stars that forms the body of the arrow itself. By placing it just on the edge or outside your finder's field of view (in the six o'clock position) you should be able to see a line of three stars near the center.

The Dumbbell Nebula is just to the south of the middle star, 14 Vulpeculae. I haven't seen it in a finder, but I've definitely snagged it with 8x30 binoculars under suburban skies. It was barely seen, circular, very misty and needed averted vision to be spotted.

If you're observing from the city, you'll probably need a telescope to see it at all. At 27x, it was seen but was extremely faint and needed averted vision. It appeared as a large, uniformly grey disk with no details seen. Increasing the magnification made observation easier.

It's much more apparent from suburban skies and should be easily seen at just 35x. It appeared rectangular at first but actually resembled a bat when observed with averted vision. Unlike the city, increasing the magnification will definitely improve the view and allow more details to be seen.

Eyepiece view courtesy *Sky Tools 3*

Albireo

Designation(s):	Beta Cygni
Constellation:	Cygnus
R.A.:	19h 30m 43s
Declination:	+27° 57' 35"
Object Type:	Multiple Star
Location:	★ ★ ★
Rating:	★ ★ ★
Best Seen:	Summer

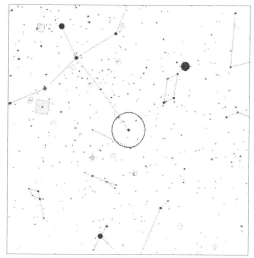

Map courtesy *Mobile Observatory*

Albireo is, by far, the most beautiful and popular double star in the entire night sky. This might sound like hyperbole, but it's really not. I challenge anyone to find a double that more easily found, easily split and with such a striking color combination.

It's not split with a regular pair of binoculars (say 10x50's) or a finderscope, but you won't need much power to reveal both components.

It's a very easy and fairly wide split at 27x with the primary being a pale gold and about 1½ times brighter than the pale blue secondary. I've noted that the strength of the colors will sometimes vary from night to night, so it's worth taking another look from time to time.

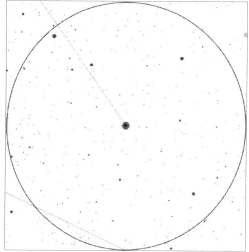

Finderscope view courtesy *Mobile Observatory*

You'll want to come back anyway – again and again – as it's a sight that's unparalleled throughout the rest of the sky. You'll show your family, your friends and your neighbors during summer barbecues and you'll anticipate their gasp – just as you did – when you first laid eyes on this beauty.

The pair lie about 430 light years away but it's not known for sure if the two components truly orbit one another. The primary, however, is itself a double but the stars are too close to be resolvable with amateur equipment.

Eyepiece view courtesy *Sky Tools 3*

Messier 29

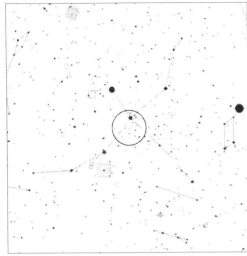
Map courtesy *Mobile Observatory*

Designation(s):	Messier 29
Constellation:	Cygnus
R.A.:	20h 23m 58s
Declination:	+38° 30′ 28″
Object Type:	Open Cluster
Location:	★
Rating:	★ ★
Best Seen:	Summer

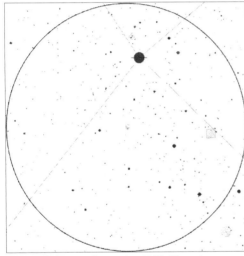
Finderscope view courtesy *Mobile Observatory*

Eyepiece view courtesy *Sky Tools 3*

Everyone should have a favorite deep sky object. Whether it's a double star, an open or globular star cluster, a nebula or a galaxy, most observers have an object they feel a particular fondness for.

Mine is Messier 29. I can't explain why. It's not particularly large or spectacular and it doesn't contain hundreds of stars. It doesn't look like diamond dust on black satin. It doesn't have any golden orange stars nestled within it.

So why bother? Well, for one thing, it's easily found as Sadr, the central star in Cygnus the Swan, is a convenient location marker for it. If you place that star about halfway between the center and the edge of your finder's view (as depicted) you should have the cluster in the center.

You may not be able to see it in the finder and I've been unable to positively identify it with binoculars, but it should be readily apparent with a small telescope and a low magnification.

My first thought, upon "discovering" this cluster, was that it looked like a mini Pleiades, which came as a surprise. It's small, is surrounded by stars and has seven or eight bright stars, all white and about the same brightness.

Given its small size, it easily fits into the field of view, even at high magnification, but I've found it best at around 100x.

Zeta Sagittae

Designation(s):	Zeta Sagittae
Constellation:	Sagitta
R.A.:	19h 48m 59s
Declination:	+19° 08' 31"
Object Type:	Multiple Star
Location:	★ ★
Rating:	★ ★
Best Seen:	Summer

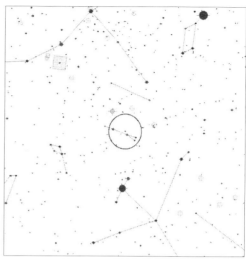

Map courtesy *Mobile Observatory*

Zeta Sagittae is a challenging double but provides a nice view for those who are able to split it. You'll need a higher power eyepiece and maybe even a barlow for this one.

Part of the challenge may also be in locating the star to begin with. It shouldn't be much of a problem if you live in the suburbs or under a dark sky as Sagitta isn't a difficult constellation to find. Just look midway between Albireo and Altair, the brightest star in Aquila the Eagle for the small, arrow shaped constellation.

If you live in a city, you'll need to use binoculars (or scan the area with your finderscope) because the chances are the stars will be too faint to see without optical aid.

Finderscope view courtesy *Mobile Observatory*

Zeta can be seen just a little to the north of the main constellation itself and close to Delta Sagittae. You won't be able to split it with low power; 35x didn't do anything for me. At 54x, the secondary was just barely visible, but unless you've already spotted it at a higher magnification, it might elude you.

Try a magnification of about 100x and then work your way down. I was able to split the pair at 91x and saw a white primary and a tiny, faint blue secondary. Now drop the power down to a lower magnification and look for the secondary again. Keep doing this until the secondary disappears – how low can you go?

Eyepiece view courtesy *Sky Tools 3*

Messier 71

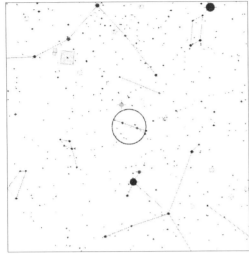
Map courtesy *Mobile Observatory*

Designation(s):	Messier 71
Constellation:	Sagitta
R.A.:	19h 53m 46s
Declination:	+18° 46' 45"
Object Type:	Globular Cluster
Location:	★
Rating:	★
Best Seen:	Summer

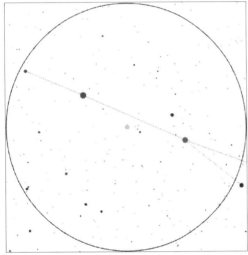
Finderscope view courtesy *Mobile Observatory*

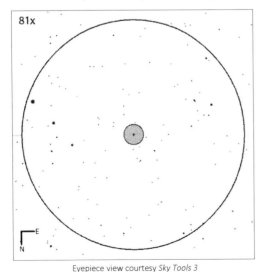
Eyepiece view courtesy *Sky Tools 3*

While you're in the vicinity of Sagitta, try your hand at Messier 71, another challenging sight for small 'scopes and located on just the other side of the constellation from Zeta.

It's too small and faint to be seen with in a finder but if you center the constellation in your field of view, the cluster should be visible through your telescope at low power.

This is a good example of an object you'd hunt down for the challenge, rather than for the eventual view. At 35x, it appears very faint, very small and very fuzzy. I've noted that it doesn't really look globular at all, but rather more like a nebula instead.

In fact, it doesn't even appear spherical. When I first observed it, I noted that it appeared rectangular and was uniformly dim with no bright core. Increasing the magnification to 70x improved the view a little but there was no resolution of the individual stars. (On the plus side, there's plenty of other stars in the field of view that will make it easy for you to focus.)

Messier 71 was discovered by Phillipe Loys de Chéseaux in 1746. The cluster lies about 27,000 light years away, is some 27 light years in diameter and was originally thought to be an open cluster.

Al Giedi

Designation(s):	Alpha Capricorni
Constellation:	Capricornus
R.A.:	20h 18m 03s
Declination:	-12° 32' 41"
Object Type:	Multiple Star
Location:	★ ★ ★
Rating:	★ ★
Best Seen:	Summer

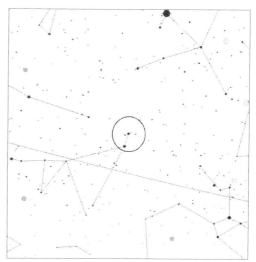

Map courtesy *Mobile Observatory*

Al Giedi is a wide, optical double, which means the two stars are not a true multiple star system.

It's a very easy object for binoculars or a finderscope with the western star appearing slightly fainter and with a hint of pale orange or cream color.

Look through a telescope and the view improves somewhat. At 35x each star will reveal a faint companion. The western star, Alpha Capricorni^2, has a wide companion while the eastern star, Alpha Capricorni^1, also has a faint companion that appears much closer to the primary star.

(Only one of these companions may be visible from the city – you may need to increase the magnification to about 80x in order to see them both. Which do you see? The wide or the close companion?)

Finderscope view courtesy *Mobile Observatory*

Al Giedi is the brightest star (or stars!) in the constellation of Capricornus, the Sea Goat. It's an ancient constellation, known to a number of cultures, with a rich mythology that goes back thousands of years.

Curiously, the constellation is almost always associated with water in some way. For example, in Greek mythology, it represents the god Pan. Part goat, the god panicked while escaping the monster Typhon and was only partly successful in turning himself into a fish before hitting the water!

Eyepiece view courtesy *Sky Tools 3*

Gamma Delphini

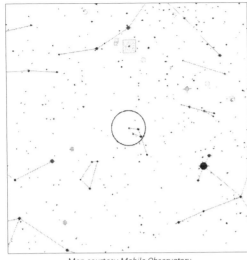
Map courtesy *Mobile Observatory*

Designation(s):	Gamma Delphini
Constellation:	Delphinus
R.A.:	20h 46m 39s
Declination:	+16° 07' 38"
Object Type:	Multiple Star
Location:	★ ★ ★
Rating:	★ ★
Best Seen:	Summer

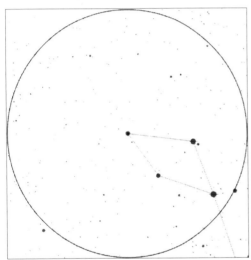
Finderscope view courtesy *Mobile Observatory*

Eyepiece view courtesy *Sky Tools 3*

Gamma Delphini is a reasonably bright star that marks one corner of a star pattern (or *asterism*) known as Job's Coffin. It's also known to be a paler version of Albireo that's an easy split for small telescopes.

Binoculars (or a finderscope) won't reveal its companion but look out for Sualocin (Alpha Delphini) just to the west and within the same field of view. You'll see a wide pair of blue-white stars with the primary about twice as bright as the secondary. (Incidentally, try spelling the name of the star backwards. It's the name of an astronomer who named it after himself!)

Gamma is easily split at 35x and is a good example of your eyes showing you color that isn't necessarily there. As explained in the notes for Lambda Arietis, two stars may appear to be of slightly different colors as a result of the brightness contrast between them.

Take a look and note their colors and then increase the magnification. Do the colors change? Maybe they will and maybe they won't. Come back on the next clear night and try it again.

I usually see a very pale yellow and very pale blue pair of stars of almost equal brightness and increasing the magnification seems to make the colors stronger. But on other nights, one looks to be pale yellow while the other is clearly white. What do you see?

Messier 15

Designation(s):	Messier 15
Constellation:	Pegasus
R.A.:	21h 29m 58s
Declination:	+12° 10' 01"
Object Type:	Open Cluster
Location:	★
Rating:	★ ★
Best Seen:	Summer and Autumn

Map courtesy *Mobile Observatory*

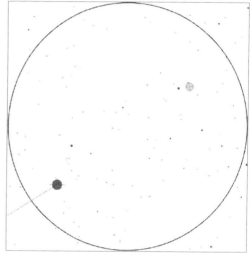

Finderscope view courtesy *Mobile Observatory*

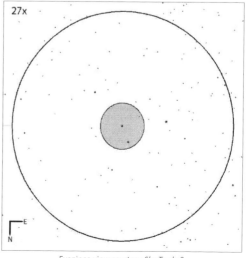

Eyepiece view courtesy *Sky Tools 3*

Messier 15 is easily found, thanks to its close proximity to Enif, the brightest star in the constellation of Pegasus, the Flying Horse.

Visible in binoculars, draw a line through Baham (to the lower left of Enif) and Enif itself and look for a faint, fuzzy star next to a slightly brighter one. It's easy to mistake it for a star but careful observation will reveal its non-stellar nature.

In a finder, with Enif in about the eight o'clock position, the cluster may appear at around the two o'clock mark. Again, look out for the star that appears close to it.

Unsurprisingly, the cluster is better observed through a telescope. At low power (around 30x) you'll see a small, hazy "star" with four real stars within the same field of view. The stars formed a zig-zag pattern and the cluster appeared between two of the stars. Averted vision may help to make the cluster appear a little larger.

The view is improved at a higher magnification. Between 80x and 100x I noticed that the core appeared surprisingly bright and extended about two thirds of the way toward the edge of the cluster. At 107x I some resolution was hinted at and the edges seemed to extend a little further out.

At about 12 billion years old, this cluster is one of the most ancient known.

Observation Logs

Date	Time	Object

_____

_____

_____

_____

_____

Date	Time	Object

_____

_____

_____

_____

_____

Date	Time	Object

_____

_____

_____

_____

_____

Date	Time	Object

_____

_____

_____

_____

_____

Date	Time	Object

_____

_____

_____

_____

_____

Date	Time	Object

_____

_____

_____

_____

_____

Date	Time	Object

_____

_____

_____

_____

_____

Date	Time	Object

_____

_____

_____

_____

_____

Date	Time	Object

_____

_____

_____

_____

_____

Date	Time	Object

_____

_____

_____

_____

_____

Date	Time	Object

_____

_____

_____

_____

_____

Date	Time	Object

_____

_____

_____

_____

_____

Date	Time	Object

_____

_____

_____

_____

Date	Time	Object

_____

_____

_____

_____

Date	Time	Object

_____

_____

_____

_____

Date	Time	Object

_____

_____

_____

_____

Date	Time	Object

_____

_____

_____

_____

Date	Time	Object

_____

_____

_____

_____

Date	Time	Object

_____

_____

_____

_____

_____

Date	Time	Object

_____

_____

_____

_____

_____

Date	Time	Object

_____

_____

_____

_____

_____

Date	Time	Object

_____

_____

_____

_____

Date	Time	Object

_____

_____

_____

_____

Date	Time	Object

_____

_____

_____

_____

Date	Time	Object

_____
_____
_____
_____
_____

Date	Time	Object

_____
_____
_____
_____
_____

Date	Time	Object

_____
_____
_____
_____
_____

Date	Time	Object

_____

_____

_____

_____

Date	Time	Object

_____

_____

_____

_____

Date	Time	Object

_____

_____

_____

_____

Date	Time	Object

_____

_____

_____

_____

_____

Date	Time	Object

_____

_____

_____

_____

_____

Date	Time	Object

_____

_____

_____

_____

_____

Date	Time	Object

_____

_____

_____

_____

_____

Date	Time	Object

_____

_____

_____

_____

_____

Date	Time	Object

_____

_____

_____

_____

_____

Date	Time	Object

_____

_____

_____

_____

_____

Date	Time	Object

_____

_____

_____

_____

_____

Date	Time	Object

_____

_____

_____

_____

_____

Date	Time	Object

_____

_____

_____

_____

_____

Date	Time	Object

_____

_____

_____

_____

_____

Date	Time	Object

_____

_____

_____

_____

_____

Date	Time	Object

_____

_____

_____

_____

_____

Date	Time	Object

_____

_____

_____

_____

_____

Date	Time	Object

_____

_____

_____

_____

_____

Date	Time	Object

_____

_____

_____

_____

_____

Date	Time	Object

_____

_____

_____

_____

_____

Date	Time	Object

_____

_____

_____

_____

_____

Date	Time	Object

_____

_____

_____

_____

_____

Date	Time	Object

_____

_____

_____

_____

_____

Date	Time	Object

_____

_____

_____

_____

_____

Date	Time	Object

_____

_____

_____

_____

_____

Date	Time	Object

_____

_____

_____

_____

_____

Date	Time	Object

_____

_____

_____

_____

_____

Date	Time	Object

_____

_____

_____

_____

_____

Date	Time	Object

_____

_____

_____

_____

_____

Date	Time	Object

_____

_____

_____

_____

_____

Date	Time	Object

_____

_____

_____

_____

_____

Date	Time	Object

_____

_____

_____

_____

_____

Date	Time	Object

_____

_____

_____

_____

_____

Date	Time	Object

_____

_____

_____

_____

_____

Date	Time	Object

_____

_____

_____

_____

_____

Date	Time	Object

_____

_____

_____

_____

_____

Date	Time	Object

_____

_____

_____

_____

Date	Time	Object

_____

_____

_____

_____

Date	Time	Object

_____

_____

_____

_____

Appendix

Deep Sky Objects by Constellation

Note: The co-ordinates have been slightly abbreviated and rounded up to the nearest minute for formatting reasons. Inputting these co-ordinates into a GoTo (with zero as the seconds) should still allow your telescope to find the object.

ANDROMEDA   Autumn and Winter

	Designation	R.A.	Dec.	Type	Location	Rating
Pi Andromedae	Pi Andromedae	00h 37m	+33° 43'	Multiple Star	★★	★★
Andromeda Galaxy	Messier 31	00h 43m	+41° 16'	Spiral Galaxy	★★	★★
NGC 752	NGC 752	01h 58m	+37° 52'	Open Cluster	★★	★★
Almach	Gamma Andromedae	02h 04m	+42° 20'	Multiple Star	★★★	★★★

ARIES   Autumn and Winter

	Designation	R.A.	Dec.	Type	Location	Rating
Mesarthim	Gamma Arietis	01h 53m	+19° 18'	Multiple Star	★★★	★★★
Lambda Arietis	Lambda Arietis	01h 58m	+23° 36'	Multiple Star	★★★	★★★

AURIGA   Winter

	Designation	R.A.	Dec.	Type	Location	Rating
Messier 37	Messier 37	05h 52m	+32° 33'	Open Cluster	★	★★★

BOÖTES   Spring and Summer

	Designation	R.A.	Dec.	Type	Location	Rating
Arcturus	Alpha Boötis	14h 16m	+19° 11'	Multiple Star	★★★	★
Delta Boötis	Delta Boötis	15h 16m	+33° 19'	Multiple Star	★★★	★★★

CANCER   Winter and Spring

	Designation	R.A.	Dec.	Type	Location	Rating
Praesepe	Messier 44	08h 40m	+19° 40'	Open Cluster	★★	★★★
Iota Cancri	Iota Cancri	08h 47m	+28° 46'	Multiple Star	★★★	★★★
Messier 67	Messier 67	08h 51m	+11° 49'	Open Cluster	★	★★

CANIS MAJOR
Winter

	Designation	R.A.	Dec.	Type	Location	Rating
Sirius	Alpha Canis Majoris	06h 45m	-16° 43'	Multiple Star	★★★	★
Messier 41	Messier 41	06h 46m	-20° 45'	Open Cluster	★★	★★★

CANIS MINOR
Winter

	Designation	R.A.	Dec.	Type	Location	Rating
Procyon	Alpha Canis Minoris	07h 39m	+05° 13'	Multiple Star	★★★	★

CANES VENATICI
Spring and Summer

	Designation	R.A.	Dec.	Type	Location	Rating
Cor Caroli	Alpha Canum Venaticorum	12h 56m	+38° 19'	Multiple Star	★★★	★★★

CAPRICORNUS
Summer

	Designation	R.A.	Dec.	Type	Location	Rating
Al Giedi	Alpha Capricorni	20h 18m	-12° 33'	Multiple Star	★★★	★★

CASSIOPEIA
Autumn and Winter

	Designation	R.A.	Dec.	Type	Location	Rating
Achird	Eta Cassiopeiae	00h 49m	+57° 49'	Multiple Star	★★★	★★★
Owl Cluster	NGC 457	01h 20m	+58° 17'	Open Cluster	★	★★★
Messier 103	Messier 103	01h 33m	+60° 39'	Open Cluster	★	★
NGC 663	NGC 663	01h 46m	+61° 13'	Open Cluster	★★	★★★

CORVUS
Spring

	Designation	R.A.	Dec.	Type	Location	Rating
Algorab	Delta Corvi	12h 30m	-16° 31'	Multiple Star	★★★	★★

CYGNUS
Summer

	Designation	R.A.	Dec.	Type	Location	Rating
Albireo	Beta Cygni	19h 31m	+27° 58'	Multiple Star	★★★	★★★
Messier 29	Messier 29	20h 24m	+38° 30'	Open Cluster	★	★★

DELPHINUS
Summer

	Designation	R.A.	Dec.	Type	Location	Rating
Gamma Delphini	Gamma Delphini	20h 47m	+16° 08'	Multiple Star	★★★	★★

DRACO
Spring and Summer

	Designation	R.A.	Dec.	Type	Location	Rating
Kuma	Nu Draconis	17h 32m	+55° 11'	Multiple Star	★★★	★★★

GEMINI
Winter

	Designation	R.A.	Dec.	Type	Location	Rating
Messier 35	Messier 35	06h 09m	+24° 21'	Open Cluster	★★	★★★
Castor	Alpha Geminorum	07h 35m	+31° 53'	Multiple Star	★★★	★★★

HERCULES
Spring and Summer

	Designation	R.A.	Dec.	Type	Location	Rating
Keystone Cluster	Messier 13	16h 42m	+36° 28'	Globular Cluster	★★	★★★
Rho Herculis	Rho Herculis	17h 24m	+37° 08'	Multiple Star	★★	★★

LEO
Spring

	Designation	R.A.	Dec.	Type	Location	Rating
Regulus	Alpha Leonis	10h 08m	+11° 58'	Multiple Star	★★★	★
Adhafera	Zeta Leonis	10h 17m	+23° 25'	Multiple Star	★★★	★★
Algieba	Gamma Leonis	10h 20m	+19° 51'	Multiple Star	★★★	★★★
Denebola	Beta Leonis	11h 49m	+14° 34'	Multiple Star	★★★	★★

LEPUS
Winter

	Designation	R.A.	Dec.	Type	Location	Rating
Gamma Leporis	Gamma Leporis	05h 44m	-22° 27'	Multiple Star	★★★	★★

LIBRA
Spring

	Designation	R.A.	Dec.	Type	Location	Rating
Zuben Elgenubi	Alpha Librae	14h 51m	-16° 00'	Multiple Star	★★★	★★

LYRA
Summer

	Designation	R.A.	Dec.	Type	Location	Rating
Double Double	Epsilon Lyrae	18h 44m	+39° 40'	Multiple Star	★★★	★★★
Sheliak	Beta Lyrae	18h 50m	+33° 22'	Multiple Star	★★★	★★
Ring Nebula	Messier 57	18h 54m	+33° 02'	Multiple Star	★	★★★

MONOCEROS
Winter

	Designation	R.A.	Dec.	Type	Location	Rating
Beta Monocerotis	Beta Monocerotis	06h 29m	-07° 02'	Multiple Star	★★	★★

ORION
Winter

	Designation	R.A.	Dec.	Type	Location	Rating
Mintaka	Delta Orionis	05h 32m	-00° 18'	Multiple Star	★★★	★★
Meissa	Lambda Orionis	05h 35m	+09° 56'	Multiple Star	★★★	★★★
Orion Nebula	Messier 42	05h 35m	-05° 23'	Nebula	★★★	★★★
Sigma Orionis	Sigma Orionis	05h 39m	-02° 36'	Multiple Star	★★★	★★★

PEGASUS
Autumn

	Designation	R.A.	Dec.	Type	Location	Rating
Messier 15	Messier 15	21h 30m	+12° 10'	Globular Cluster	★	★★

PERSEUS

Autumn and Winter

	Designation	R.A.	Dec.	Type	Location	Rating
Double Cluster	NGC 869 & 884	02h 22m	+57° 09'	Open Clusters	★★★	★★★
Messier 34	Messier 34	02h 42m	+42° 45'	Open Cluster	★★	★★

PUPPIS

Winter

	Designation	R.A.	Dec.	Type	Location	Rating
Messier 93	Messier 93	07h 44m	-23° 51'	Open Cluster	★	★★

SAGITTA

Summer

	Designation	R.A.	Dec.	Type	Location	Rating
Zeta Sagittae	Zeta Sagittae	19h 49m	+19° 09'	Multiple Star	★★	★★
Messier 71	Messier 71	19h 54m	+18° 47'	Globular Cluster	★	★

SAGITTARIUS

Summer

	Designation	R.A.	Dec.	Type	Location	Rating
Messier 22	Messier 22	18h 36m	-23° 54'	Globular Cluster	★★	★★

SCORPIUS

Summer

	Designation	R.A.	Dec.	Type	Location	Rating
Graffias	Beta Scorpii	16h 05m	-19° 48'	Multiple Star	★★★	★★★
Jabbah	Nu Scorpii	16h 12m	-19° 28'	Multiple Star	★★	★★★
Messier 80	Messier 80	16h 17m	-22° 59'	Globular Cluster	★	★
Messier 4	Messier 4	16h 24m	-26° 32'	Globular Cluster	★	★★
Butterfly Cluster	Messier 6	17h 40m	-32° 15'	Open Cluster	★★	★★★
Messier 7	Messier 7	17h 54m	-34° 48'	Open Cluster	★★	★★

SCUTUM

Summer

	Designation	R.A.	Dec.	Type	Location	Rating
Wild Duck Cluster	Messier 11	18h 51m	-06° 16'	Open Cluster	★★	★★

TAURUS
Autumn and Winter

	Designation	R.A.	Dec.	Type	Location	Rating
Pleiades	Messier 45	03h 48m	+24° 10'	Open Cluster	★★★	★★★
Crab Nebula	Messier 1	05h 35m	+22° 01'	Supernova Remnant	★	★

URSA MAJOR
Spring and Summer

	Designation	R.A.	Dec.	Type	Location	Rating
Mizar & Alcor	Zeta and 80 Ursae Majoris	13h 24m	+54° 56'	Multiple Star	★★★	★★★

URSA MINOR
All Year Round

	Designation	R.A.	Dec.	Type	Location	Rating
Polaris	Alpha Ursae Minoris	02h 32m	+89° 16'	Multiple Star	★★★	★

VULPECULA
Summer

	Designation	R.A.	Dec.	Type	Location	Rating
Coathanger Cluster	Collinder 399	19h 26m	+20° 13'	Open Cluster	★★	★★
Dumbbell Nebula	Messier 27	20h 00m	+22° 43'	Planetary Nebula	★	★★

Greek Alphabet

Alpha	α	Epsilon	ε	Iota	ι	Nu	ν	Rho	ρ	Phi	φ
Beta	β	Zeta	ζ	Kappa	κ	Xi	ξ	Sigma	σ	Chi	χ
Gamma	γ	Eta	η	Lambda	λ	Omicron	ο	Tau	τ	Psi	ψ
Delta	δ	Theta	θ	Mu	μ	Pi	π	Upsilon	υ	Omega	ω

Elongations of Mercury and Venus, 2016-2025

	Mercury		Venus	
2016	February 6th	Morning Sky		
	April 17th	Evening Sky		
	June 4th	Morning Sky		
	August 16th	Evening Sky		
	September 28th	Morning Sky		
	December 10th	Evening Sky		
2017	January 18th	Morning Sky	January 11th	Evening Sky
	March 31st	Evening Sky	June 2nd	Morning Sky
	May 17th	Morning Sky		
	July 29th	Evening Sky		
	September 11th	Morning Sky		
	November 23rd	Evening Sky		
2018	January 1st	Morning Sky	August 16th	Evening Sky
	March 14th	Evening Sky		
	April 28th	Morning Sky		
	July 11th	Evening Sky		
	August 26th	Morning Sky		
	November 5th	Evening Sky		
	December 14th	Morning Sky		
2019	February 26th	Evening Sky	January 5th	Morning Sky
	April 11th	Morning Sky		
	June 23rd	Evening Sky		
	August 9th	Morning Sky		
	October 19th	Evening Sky		
	November 27th	Morning Sky		
2020	February 9th	Evening Sky	March 24th	Evening Sky
	March 23rd	Morning Sky	August 12th	Morning Sky
	June 3rd	Evening Sky		
	July 21st	Morning Sky		
	September 30th	Evening Sky		
	November 9th	Morning Sky		
2021	January 23rd	Evening Sky	October 29th	Evening Sky
	March 5th	Morning Sky		
	May 16th	Evening Sky		
	July 4th	Morning Sky		
	September 13th	Evening Sky		
	October 24th	Morning Sky		

Elongations of Mercury and Venus, 2016-2025 (cont.)

	Mercury		Venus	
2022	January 6th	Evening Sky	March 19th	Morning Sky
	February 16th	Morning Sky		
	April 28th	Evening Sky		
	June 15th	Morning Sky		
	August 26th	Evening Sky		
	October 8th	Morning Sky		
	December 20th	Evening Sky		
2023	January 29th	Morning Sky	June 3rd	Evening Sky
	April 11th	Evening Sky	October 23rd	Morning Sky
	May 28th	Morning Sky		
	August 9th	Evening Sky		
	September 21st	Morning Sky		
	December 3rd	Evening Sky		
2024	January 11th	Morning Sky		
	March 24th	Evening Sky		
	May 9th	Morning Sky		
	July 21st	Evening Sky		
	September 4th	Morning Sky		
	November 15th	Evening Sky		
	December 24th	Morning Sky		
2025	March 7th	Evening Sky	January 9th	Evening Sky
	April 20th	Morning Sky	May 31st	Morning Sky
	July 3rd	Evening Sky		
	August 18th	Morning Sky		
	October 29th	Evening Sky		
	December 7th	Morning Sky		

Oppositions of Mars, Jupiter and Saturn, 2016-2025

	Mars		Jupiter		Saturn	
2016	May 21st	Scorpius	March 7th	Leo	June 2nd	Ophiuchus
2017			April 7th	Virgo	June 14th	Ophiuchus
2018	July 26th	Capricornus	May 8th	Libra	June 27th	Sagittarius
2019			June 9th	Ophiuchus	July 8th	Sagittarius
2020	October 13th	Pisces	July 14th	Sagittarius	July 20th	Sagittarius
2021			August 19th	Capricornus	August 1st	Capricornus
2022	December 7th	Taurus	September 26th	Pisces	August 13th	Capricornus
2023			November 2nd	Aries	August 27th	Aquarius
2024			December 7th	Taurus	September 7th	Aquarius
2025	January 15th	Gemini			September 20th	Pisces

Complete List of Constellations

Latin	Genitive	English Name	Abbreviation	Size
Andromeda	Andromedae	The Princess	And	19th
Antila	Antilae	The Air Pump	Ant	62nd
Apus	Apodis	The Bird of Paradise	Aps	67th
Aquarius	Aquarii	The Water Bearer	Aqr	10th
Aquila	Aquilae	The Eagle	Aql	22nd
Ara	Arae	The Altar	Ara	63rd
Aries	Arietis	The Ram	Ari	39th
Auriga	Aurigae	The Charioteer	Aur	21st
Boötes	Boötis	The Herdsman	Boo	13th
Caelum	Caeli	The Chisel	Cae	81st
Camelopardalis	Camelopardalis	The Giraffe	Cam	18th
Cancer	Cancri	The Crab	Cnc	31st
Canes Venatici	Canum Venaticorum	The Hunting Dogs	CVn	38th
Canis Major	Canis Majoris	The Large Dog	CMa	43rd
Canis Minor	Canis Minoris	The Little Dog	CMi	71st
Capricornus	Capricorni	The Sea Goat	Cap	40th
Carina	Carinae	The Keel	Car	34th
Cassiopeia	Cassiopeiae	The Queen	Cas	25th
Centaurus	Centauri	The Centaur	Cen	9th
Cepheus	Cephei	The King	Cep	27th
Cetus	Ceti	The Sea Monster	Cet	4th
Chamaeleon	Chamaeleontis	The Chameleon	Cha	79th
Circinus	Circini	The Compass	Cir	85th
Columba	Columbae	The Dove	Col	54th
Coma Berenices	Comae Berenices	Berenices' Hair	Com	42nd
Corona Australis	Coronae Australis	The Southern Crown	CrA	80th
Corona Borealis	Coronae Borealis	The Northern Crown	CrB	73rd
Corvus	Corvi	The Crow	Crv	70th
Crater	Crateris	The Cup	Crt	53rd
Crux	Crucis	The Southern Cross	Cru	88th
Cygnus	Cygni	The Swan	Cyg	16th
Delphinus	Delphini	The Dolphin	Del	69th
Dorado	Doradus	The Goldfish	Dor	72nd
Draco	Draconis	The Dragon	Dra	8th
Equuleus	Equulei	The Foal	Equ	87th
Eridanus	Eridani	The River	Eri	6th

Complete List of Constellations (cont.)

Latin	Genitive	English Name	Abbreviation	Size
Fornax	Fornacis	The Furnace	For	41st
Gemini	Geminorum	The Twins	Gem	30th
Grus	Gruis	The Crane	Gru	45th
Hercules	Herculis	The Hero	Her	5th
Horologium	Horologii	The Clock	Hor	58th
Hydra	Hydrae	The Water Snake	Hya	1st
Hydrus	Hydri	The Lesser Water Snake	Hyi	61st
Indus	Indi	The Indian	Ind	49th
Lacerta	Lacertae	The Lizard	Lac	68th
Leo	Leonis	The Lion	Leo	12th
Leo Minor	Leonis Minoris	The Little Lion	LMi	64th
Lepus	Leporis	The Hare	Lep	51st
Libra	Librae	The Scales	Lib	29th
Lupus	Lupi	The Wolf	Lup	46th
Lynx	Lyncis	The Lynx	Lyn	28th
Lyra	Lyrae	The Lyre	Lyr	52nd
Mensa	Mensae	The Table Mountain	Men	75th
Microscopium	Microscopii	The Microscope	Mic	66th
Monoceros	Monocerotis	The Unicorn	Mon	35th
Musca	Muscae	The Fly	Mus	77th
Norma	Normae	The Carpenter's Level	Nor	74th
Octans	Octantis	The Octant	Oct	50th
Ophiuchus	Ophiuchi	The Serpent Bearer	Oph	11th
Orion	Orionis	The Hunter	Ori	26th
Pavo	Pavonis	The Peacock	Pav	44th
Pegasus	Pegasi	The Flying Horse	Peg	7th
Perseus	Persei	The Hero	Per	24th
Phoenix	Phoenicis	The Phoenix	Phe	37th
Pictor	Pictoris	The Painter's Easel	Pic	59th
Pisces	Piscium	The Fishes	Psc	14th
Piscis Austrinus	Piscis Austrini	The Southern Fish	PsA	60th
Puppis	Puppis	The Poop Deck	Pup	20th
Pyxis	Pyxidis	The Compass	Pyx	65th
Reticulum	Reticuli	The Net	Ret	82nd

Complete List of Constellations (cont.)

Latin	Genitive	English Name	Abbreviation	Size
Sagitta	Sagittae	The Arrow	Sge	86th
Sagittarius	Sagittarii	The Archer	Sgr	15th
Scorpius	Scorpii	The Scorpion	Sco	33rd
Sculptor	Sculptoris	The Sculptor	Scl	36th
Scutum	Scuti	The Shield	Sct	84th
Serpens	Serpentis	The Serpent	Ser	23rd
Sextans	Sextantis	The Sextant	Sex	47th
Taurus	Tauri	The Bull	Tau	17th
Telescopium	Telescopii	The Telescope	Tel	57th
Triangulum	Trianguli	The Triangle	Tri	78th
Triangulum Australe	Trianguli Australis	The Southern Triangle	TrA	83rd
Tucana	Tucanae	The Toucan	Tuc	48th
Ursa Major	Ursae Majoris	The Great Bear	UMa	3rd
Ursa Minor	Ursae Minoris	The Little Bear	UMi	56th
Vela	Velorum	The Sails	Vel	32nd
Virgo	Virginis	The Virgin	Vir	2nd
Volans	Volantis	The Flying Fish	Vol	76th
Vulpecula	Vulpeculae	The Fox	Vul	55th

Glossary

Aperture

The aperture of a telescope is the width of the lens on a refractor telescope or the width of the open end of a reflector telescope tube. In both cases, this is the end that points up toward the sky.

See also *Reflector Telescope* and *Refractor Telescope*

Astronomical Unit

One Astronomical Unit (AU) is the distance from the Earth to the Sun. This is taken as 149,597,870 kilometers or 92,955,806 miles. To make life (and polite conversation as your astronomical society) easier, this is taken as 150 million kilometers or 93 million miles.

See also *Light Year*

Averted Vision

Averted vision is the "trick" of looking at an object out of the corner of your eye. As your eyes are more sensitive to light (rather than color) with averted vision, you're more likely to see a particularly faint object. Averted vision can be particularly useful in observing the details in a nebula.

Barlow Lens

A barlow lens is an accessory that allows you to effectively double (or, with some, triple) the magnification of any eyepiece used with it. A barlow is considered by many to be an essential tool for the amateur astronomer.

Declination

Just as geographical co-ordinates use longitude and latitude, stellar cartographers use right ascension and declination, respectively. Declination, like latitude, is measured in degrees and indicates an object's position in the sky relative to the northern and southern hemisphere. For example, Polaris has a declination of almost +90° and is therefore almost overhead from the North Pole. An object with a declination of 0° will pass overhead when observed from the equator. There are no bright stars with a declination of -90° and, therefore, no star marking the celestial South Pole.

See also *R.A. (Right Ascension)*

Focal Length

Focal length is the distance that light travels from the telescope's entry point (i.e., the lens on a refractor telescope or the aperture on a reflector) to the eyepiece. It is measured in millimeters. The longer the focal length, the higher the possible magnification of the telescope.

See also *Aperture, Reflector Telescope* and *Refractor Telescope.*

Globular Clusters

There are two types of star cluster: open and globular. Globular star clusters are like huge balls of stars, with each one containing thousands of stars. These stars are packed into a relatively small space, often with only a few light years between each one. The clusters form a halo around the galactic center and mostly contain older stars, often billions of years old.

See also *Open Clusters*

Greatest Eastern/Western Elongation

As both Mercury and Venus orbit the Sun at a closer distance than the Earth, they never stray very far from the Sun in the sky. However, there comes a time when each planet has moved the furthest it can before it starts to move back towards the Sun again. This point is called its greatest elongation.

Greatest eastern elongation means the planet is at its furthest point to the east of the Sun, thereby making it visible in the evening sky (confusingly, in the *west* after sunset.) When the planet is at its greatest western elongation, it's visible in the morning sky in the *east* before sunrise.

See also *Opposition*

Light Year

A light year is the distance that light travels in a year. Light travels at an approximate speed of 186,000 mph per *second*. In other words, at a distance of roughly 92 million miles, light from the Sun takes about eight minutes to reach us.

Given that the distance from the Earth to the Sun is one *Astronomical Unit*, it can be said that light travels 7 ½ AUs every hour, 180 AUs every day, 1,260 AUs every week, about 5,400 every month and 65,520 every year. The closest star is Proxima Centauri, which, at a distance of 4.24 light years, is some 277,804 AUs away. (And this is why we don't measure every distance in miles and kilometers.)

See also *Astronomical Unit*

Nebula (pl. nebulae)

A nebula is a huge cloud of gas and dust in space. These clouds are light years in diameter and are the birthplaces of the stars themselves. As the gas and dust particles move through space, they collide with other particles and gradually begin to grow in mass. Eventually, the clump of gas becomes so massive that a gravitational collapse begins, causing nuclear fusion and the birth of a star. Most of the nebulae are located away from bright stars (and may require a little practice to locate) but the Orion Nebula is easily seen with unaided eyes on a winter's night.

At the other end of a star's life are planetary nebulae. A planetary nebula is the last dying gasp of a star. It's the shell of gas and dust that the star ejects as it collapses and dies, causing the shell to expand out into space. They're named *planetary* nebulae because many look like the tiny disc of a planet when observed through a telescope.

The most famous planetary nebula is the Ring Nebula, which looks just like a smoke ring in space. This can easily be seen with a small telescope but you might need a slightly larger 'scope (around 100mm or more) to see the central hole itself. You can see the Ring Nebula during the summer months.

If the star goes supernova, you may well be able to see the remnants in the night sky. Many of these are faint (such as the Veil Nebula, which actually looks like the remains of an exploded star) but one, the Crab Nebula, should also be visible on a winter's night.

Open Clusters

Open clusters are a type of star cluster. Whereas globular clusters appear small, spherical and often faint, most open clusters are quite the opposite. Many contain tens or hundreds of stars of varying brightness, scattered across an area of sky sometimes larger than the full Moon. There are a number that are easily observed in a small telescope with the Pleiades, Praesepe and Butterfly clusters being fine examples.

See also *Globular Clusters*

Opposition

A planet (or other solar system body, such as an asteroid) is said to be at opposition when it's directly opposite the Sun in the sky. When this occurs, it rises at sunset and sets at sunrise and is therefore visible throughout the entire night. This is when the planet is at its best for observation, as it will be at its closest to the Earth and will appear largest and brightest in the night sky.

Opposition can only occur if the body orbits the Sun at a greater distance than the Earth. For example, the planets Mars, Jupiter and Saturn all periodically reach opposition but the planets Mercury and Venus will always appear close to the Sun in the sky.

See also *Greatest Eastern/Western Elongation*

Optical Binary

An optical binary is a chance alignment of two stars in the sky, thereby giving the illusion that the pair are a true multiple star system. In reality, the stars may be hundreds of light years apart and moving in different directions through space.

See also *Primary Component* and *Secondary Component*

Planetary Nebulae

See *Nebula*

Primary Component

The primary component of a multiple star system is the brightest with the next brightest being the secondary. For example, the famous double star Albireo has a golden primary with a sapphire blue secondary.

See also *Optical Binary* and *Secondary component*

R.A. (Right Ascension)

On star maps and charts, R.A. is the celestial equivalent of longitude and enables astronomers to easily locate an object in the night sky. Unlike longitude, it's measured in hours, minutes and seconds with 00h 00m 00s holding a special significance. When the Sun reaches that point in its movement across the sky, the March (vernal) equinox occurs. Spring then begins in the northern hemisphere and while autumn begins in the south.

See also *Declination*

Reflector Telescope

A reflector telescope is one that works using mirrors. The light enters through the open end (the *aperture*) and then hits the primary mirror at the bottom of the telescope tube. The light is then reflected back up towards the aperture, where it hits a smaller secondary mirror and is directed into the eyepiece. This has the benefit of having the eyepiece conveniently located near the top of the tube and also allows for a longer *focal length* without increasing the length of the tube itself.

See also *Aperture, Focal Length* and *Refractor Telescope.*

Refractor Telescope

A refractor telescope uses lenses to magnify the image for the observer. It comprises of a long tube with a primary lens at the aperture, where the light enters the telescope. The light then travels all the way down the tube and then exits through the eyepiece at the bottom.

See also *Aperture, Focal Length* and *Reflector Telescope.*

Secondary Component

The secondary component of a multiple star system is typically the second brightest star, with the brightest being the primary.

See also *Optical Binary* and *Primary Component*

Supernova Remnant

See *Nebula.*

Recommended Resources

Facebook Groups
- *Astronomy for Beginners (A4B)*
- *Astronomy for Fun*
- *Astronomy Workfile*
- *Online Astronomy Society*
- *Space Science and Astronomy for Home Educators*
- *Telescope Addicts – Astronomy & Astrophotography*
- *UK Astronomy*

Books
- *365 Starry Nights* – Chet Raymo
- *Astronomy Hacks* – Robert Bruce Thompson & Barbara Fritchman Thompson
- *Binocular Highlights* – Gary Seronik
- *Celestial Harvest* – James Mullaney
- *Celestial Sampler* – Sue French
- *Cosmic Challenge* – Philip S. Harrington
- *Deep Sky Observing with Small Telescopes* – David J. Eicher
- *Deep-Sky Wonders* – Sue French
- *Deep Sky Wonders* – Walter Scott Houston
- *Double Stars for Small Telescopes* – Sissy Haas
- *Illustrated Guide to Astronomical Wonders* – Robert Bruce Thompson & Barbara Fritchman Thompson
- *Observer's Sky Atlas, The* – E. Karkoschka
- *Observing Handbook and Catalogue of Deep-Sky Objects* – Christian B. Luginbuhl and Brian A. Skiff
- *Planet Observer's Handbook, The* – Fred W. Price
- *Pocket Sky Atlas* – Roger W. Sinnott
- *Star-Hopping for Backyard Astronomers* – Alan M. MacRobert
- *Star Names – Their Lore and Meaning* – Richard Hinckley Allen
- *Star Watch* - Philip S. Harrington
- *Touring the Universe Through Binoculars* – Phillip S. Harrington
- *Turn Left at Orion* – Guy Consolmagno and Dan M. Davis

Software
- *Mobile Observatory* by Wolfgang Zima (http://zima.co.)
- *Sky Tools* by Greg Crinklaw (http://www.skyhound.com)
- *Stellarium* (Open Source software – http://www.stellarium.org)

Made in the USA
Middletown, DE
29 November 2018